I0475470

PHILOSOPHY AND THE
NEW PHYSICS

ROUGIER

PHILOSOPHY AND THE NEW PHYSICS

An Essay on the Relativity Theory and the Theory of Quanta

BY

LOUIS ROUGIER

PROFESSEUR AGRÉGÉ DE PHILOSOPHIE, DOCTEUR ÈS LETTRES

Authorized Translation
From the Author's Corrected Text of
'La Matérialisation de l'Énergie"

BY

MORTON MASIUS, M.A., PH.D.

PROFESSOR OF PHYSICS IN THE WORCESTER POLYTECHNIC INSTITUTE
TRANSLATOR OF M. PLANCK'S THEORY OF RADIATION

PHILADELPHIA
P. BLAKISTON'S SON & CO.
1012 WALNUT STREET

QC 6
R 68

COPYRIGHT, 1921, BY P. BLAKISTON'S SON & CO.

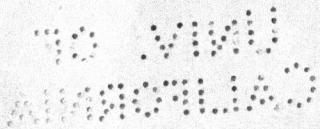

UNIV. OF
CALIFORNIA

THE MAPLE PRESS YORK PA

TRANSLATOR'S PREFACE

The recent remarkable developments of physical theories, especially those concerned with relativity and quanta of energy, cannot fail to have far-reaching influences on philosophical thought. Physicists, as a rule, are too much occupied with their special field to give much attention to matters of more general philosophical interest, and few philosophers possess the knowledge of science required for discussing and criticizing fruitfully the work of the physicist. Professor Rougier's very wide reading in mathematical and experimental Physics has enabled him to present and interpret the new advances in Physics in a way which should prove of great interest to both philosopher and physicist. This book seems to mark a measurable advance toward a confluence of the broad streams of philosophical and scientific enquiry.

M. M.

449830

CONTENTS

INTRODUCTION

It has been customary since the time of Auguste Comte (1795–1857) to distinguish two entirely separate categories of problems: those which may be treated by the methods of science and which are, because of the perfection these methods have reached, capable of being solved sooner or later, and those which, being beyond these methods, lie outside of the limits of experience and on that account are called metaphysical problems. The former problems, taken together, mark off the field of the exact science; the latter comprise what Herbert Spencer and Du Bois-Reymond have called the unknowable.

We have no criterion by means of which to decide *a priori* to which of these two categories a given problem belongs. In every case in which a philosophical system has claimed to lay down, dogmatically, the limits to our experience, later scientific discoveries have seemed to make a point of proving it to be wrong. For example, the real nature of phenomena, according to Auguste Comte, the founder of positivism, will never be disclosed; "When a scientific theory," declares Poincaré,[1]

[1] H. Poincaré, La valeur de la Science, p. 267.

ix

"claims to tell us what heat, what electricity, or what life really is, it stands convicted at the outset." This idea led Ostwald to his *Energetics*, and Duhem to his *Theory of Physics*. Who at the present time would bind himself to this prudent agnosticism, to the extent of doubting the objective existence of discontinuous elements, like molecules, atoms, or electrons? "The atoms are no longer a convenient fiction; it seems to us that we can, so to speak, see them, since we know how to count them,"[1] as Poincaré had to admit in the later part of his life when confronted by the achievements of the atomic hypotheses.

It occurs much more often that metaphysical problems, supposedly incapable of solution, vanish, simply because the progress of ideas shows that they are fictitious problems or pseudo-problems, or problems which have been badly conceived. We have nothing but pity mingled with boredom for the endless disputes of the Scholastics on the subject of the unity or the plurality of the substantial forms in animal species. We know, in fact, that nothing but individuals really exist, and that, to the static abstractions of our mind regarding these genera and species, nothing invariable and essential corresponds, beyond a mere bond of relationship and a family resemblance between the individuals that we range in the same class. The question of knowing whether a body is at rest or in absolute motion is, according to Einstein, another example

[1] H. Poincaré, Dernières pensées, p. 196.

of a pseudo-problem. There exists no absolute space which might serve as a privileged reference system; there exist merely bodies at rest or in motion relatively to one another.

It is a metaphysical problem of the same kind that arises from the fundamental dualism between ponderable matter and imponderable energy, which classical physics uses for the basis of an explanation of the world and which gives place to two principles of invariance, the principle of the conservation of mass and the principle of the conservation of energy. There is a radical difference of character between these two components of all phenomena of nature, matter and energy; matter alone is endowed with mass, with weight in proportion thereto, and with structure; energy has no inertia, no weight, and no structure. Therefore how is it to be conceived that an imponderable noncorpuscular agent that has no distribution in space capable of representation can be applied to a body with both mass and inertia, and possessing a definite shape, and act on it to the point of deforming or moving it? How, for example, can luminous radiation, if it is destitute of mass, exert a repulsion in one direction on the source which emits it and an impulse on the opaque body absorbing it, as if it represented momentum; how, in other words, if it is deprived of inertia, can it behave like a material projectile which exerts a recoil pressure on the fire-arm which throws it and a ballistic action on the obstacle which it strikes? And if energy possesses some inertia how can it accumulate without effect on a body, like electric energy, without the mass of this body increasing? This is the metaphysical problem of the mutual action of energy and matter. If one assimilates mental labor to a kind of energy *sui generis*, the metaphysical problem of the relations between mind and body presents itself as a special case of the foregoing.

Vain attempts have been made to reduce one of the two terms placed in contrast to the other by the dualistic theory, in order to escape the problem of their mutual action; these monistic attempts have merely served to shift the problem by substituting new difficulties for the one that they attempted to overcome.

All this became quite different as soon as the theory of relativity of Einstein and the theory of *quanta* of Max Planck led, if not to a complete rejection of the fundamental dualism of matter and energy, at any rate to an approach of the two terms by assigning to them such common properties as to render their relations intelligible. Thus these theories confer on energy inertia, weight in proportion thereto, and even a certain kind of structure. Consequently, radiation represents electromagnetic momentum and may be properly likened to a material projectile. The pressure of radiation ceases to be incomprehensible and to give rise to the vexatious metaphysical problem of the action of the imponderable on the ponderable.

Not only does the inertia of energy, which as late

as 1913 Brillouin[1] described as a "paradoxical fancy," relieve us from the drawbacks of such a problem, but it satisfies our inherent need of unification by absorbing the principle of the conservation of mass into the more general one of the conservation of energy and by extending the law of Newtonian attraction to radiant energy. Moreover the fundamental formula $E = mV^2$ leads to the following important result: it permits the evaluation of the internal energy of a gram of matter at the absolute zero for an observer at rest, an energy which is no less than that of 3 million kilograms of coal.

The weight of energy, which, according to the experiments of Eötvös, is proportional to its inertia, furnishes, as Langevin has shown, a natural interpretation of the discrepancies in Prout's law of atomic weights. It is the origin of the theory of gravitation of Einstein based on the generalized principle of relativity. The practical interest of this theory is that it leads to a successful calculation of the secular anomaly of the perihelion of Mercury, and to the correct interpretation of the shift of Fraunhofer's lines in the solar spectrum as compared with those from a terrestrial source, observed by Fabry and Buisson. But its essential advantage lies in the fact that it excludes from the domain of physics those metaphysical entities that are still encumbering it, such as the absolute space and the

---

[1] Brillouin, Propos sceptiques au sujet du principe de relativité (Scientia, janvier 1913, p. 23).

privileged axes of Newton and their successors, the body *alpha* of Neumann and the stationary ether of the opticians. Psycho-physiology teaches us with respect to this point that our senses perceive merely relative variations in the external world without ever detecting an absolute change. Einstein has reconciled natural philosophy with the requirements of epistemology by showing that it is possible to put the equations of physics into a form which is generally invariant with respect to all changes of coordinate axes. By making use of the absolute differential calculus he has shown that it is thus possible to substitute for coordinate equations intrinsic equations expressed in terms of tensor equalities in order to have in the terminology of the physical laws the magnitudes characteristic of the gravitational field; or, putting it more exactly, to regard the physical laws as relations between the quantities characteristic of the gravitational field and the quantities characteristic of the special phenomena to be studied.

Lastly, free radiation seems to possess structure. The law of energy distribution in the spectrum of the black body and the study of the specific heats of solids at low temperatures lead to the belief that the energy exchanges between material systems by means of radiation take place in sudden jumps, according to integral multiples of elementary indivisible quantities, veritable energy atoms, called *quanta*. The discontinuity of emission and absorption of radiation paves the way for the idea of a discontinuous distribution of radiant energy in the front of transverse light waves. This follows necessarily from the interpretation of certain phenomena such as the photo-electric effect and the production of rays by the impulse of secondary cathode rays or X-rays. Consequently radiation appears no longer to be a form of energy propagated after the manner of continuous' waves through a hypothetical stationary medium, Maxwell's dielectric ether, but as expelled into space in the form of discrete units with a uniform velocity in a constant gravity field. The ancient dualism of the ponderable and the imponderable, of matter and energy, becomes transformed into one of energy stabilized in material structures of definite architecture and free radiation, both of these modes of energy being endowed equally with inertia, weight in proportion thereto, and structure. This may then be called the materialization of energy.

Philosophy and the New Physics

CHAPTER I

THE DUALISM OF MATTER AND ENERGY

1. INTRODUCTION.

It is a general truth that the majority of philosophical problems are insoluble because the problems do not properly exist. The subjectivism of our senses, the anthropomorphism of our reasoning by analogy, the substantialistic tendency to realize our ideas and to take purely logical distinctions as objects lead us to conceive fictitious problems, or pseudo-problems, that have no more meaning than the *insolubilia* on which the eristics of the ancient sophists or the forensic dialectics of the theological schools of the middle ages were exercised. To solve them is always to show that they were problems which have been badly stated.

One category of these pseudo-problems is derived from the mental transformation of a simple analogy into an absolute identity, or of a partial difference into a perfect contrast. Having obtained certain dissimilarities between two classes of phenomena, we deliberately deny to those of one class everything that is shown to belong to those of the other; and in this perfect antithesis the mind derives the satisfaction of symmetry. We hold this radical line of demarkation to be the equivalent of reality. If we begin to reflect on the fact that these diametrically opposed phenomena show mutual interaction, then the shock of this possibility outrages our understanding; how can absolutely heterogeneous agents be combined, and how can they influence each other? For example, having supposed as a fact that matter alone has mass, proportional weight, and shape; and that force or energy possesses no inertia, no weight, and no structure, then the following problem inevitably arises: How can an imponderable agent be applied to and act on a ponderable one, with mass and inertia, to the point of deforming it or imparting an impulse to it? In accordance with the adage of Leibniz, "*causam aequat effectus*," our mind refuses to conceive this and our imagination declares itself incapable of visualizing it. Of the same nature and issuing from the same origin is the problem of the relations between mind and body.

The solutions proposed for such problems inevitably reiterate the following theme. In a *dogmatic period* it is declared that one of the two terms of the antithesis exists only apparently, the other one alone being reality; one tries to reduce the real forces either to thrusts of elementary masses or to fictitious forces of connection and inertia, or, inversely, one sacrifices the notion of material mass in order to avoid contemplating anything but force centers that attract or repel in a straight line, according to a certain power of the distance. In supporting such reductions one perceives that insurmountable difficulties which merely displace the original antinomy are encountered. The dogmatic period is followed by the prudent agnosticism of a *positive period*. It is maintained that science is merely descriptive, and one is content to express the laws that govern the relations of the two classes of agents without seeking to penetrate into their nature or the hidden mechanism of their interaction; this is, in the case in which we are interested, the period of Ostwald's energetics. Finally an *experimental and critical period* arrives; in it the claims of the two terms under discussion to be placed in antithesis are examined, and it is then discovered that the latter is not well founded. On the contrary, it is shown to be true that the two terms, taken to be diametrically opposite, enjoy such properties in common as explain their interaction; energy appears to be endowed with inertia, weight, and structure, like matter. The profound reason for the lack of success of the preceding attempts at reduction is accounted for, and the metaphysical problem, thus removed unexpectedly, vanishes of itself.

2. THE DUALISTIC THEORY.

A superficial view of the external world leads to a classification of the agents found in it into two quite distinct categories. In the first place, there are ponderable bodies, endowed with mass, weight, and structure, the aggregate of which constitutes matter; in the second place, there are imponderable forces, divested of inertia, weight, and structure, the aggregate of which constitutes energy. Take a piece of matter and make it undergo all the possible physical and chemical transformations, such as motions; subdivisions and recompositions; expansions and compressions; electrification; magnetization; changes of state; and chemical combinations. There exists an invariant with respect to this group of physicochemical transformations; that is, a certain coefficient characteristic of the individuality of this piece of matter which remains invariable and unchanged: this is its mass. The mass, therefore, serves for measuring a portion of matter so well that matter and mass appear to be synonyms. Lavoisier established the indestructibility of matter by showing, by means of accurate weighing, the conservation of mass. A certain amount of energy is, because of the principle that different forms of energy are equivalent, measured by the amount of mechanical work into which it may be converted; and experiment proves that, in an isolated system, the amount of energy is constant. This is summarized by a classical passage of Robert Mayer:[1] "Nature presents two categories of agents between which experiment shows an insurmountable barrier to exist. The first category comprises agents having the properties of being ponderable and impenetrable: these are forms of matter; the second comprises agents lacking these properties: these are forces

[1] Robert Mayer, Annalen der Pharmacie und Chemie, von Liebig und Wöhler, 1842.

called imponderables on account of the negative property which characterizes them. Forces are indestructible, variable and imponderable objects."

Matter and energy being equally indestructible and radically heterogeneous, neither can grow at the expense of the other or decrease to its gain. Matter must return the energy to an amount equivalent to what has been given to it in another form. It serves to store it up like a sponge that has been saturated with water and is then in turn pressed dry; but, in contrast with the analogous case, the presence of energy in matter does not increase its mass since energy is imponderable. Matter is the natural receptacle of energy, which does not exist independently of it. Mass of ponderable bodies and imponderable energy are conserved, each on its own account like two distinct worlds, one superimposed on the other, which completely penetrate without knowing each other and without consenting to reciprocal exchanges. This is what a recent author expresses as follows:[1]

"The world where we live is in reality a double world, or rather it consists of two distinct worlds, one of which is the world of matter, the other the world of energy. Copper, iron, carbon, that is the world of matter. Mechanical work, heat, these are forms of energy. Each of these two worlds is governed by a law of conservation. Matter can not be created nor destroyed; energy can not be created nor destroyed."

[1] Paul Janet, Leçons d'Électricité, 2 édition, pp. 2 and 5.

"Matter or energy may appear in a large number of forms, without matter ever changing into energy or energy into matter."

"We can no more conceive of energy without matter than of matter without energy."

This dualistic doctrine of the universe rests on the following principle:

Matter alone is endowed with mass, weight, and structure; energy has no mass, no weight, and no structure.

This principle is supposed to be founded on the following propositions which we shall hereafter call the postulates of the theory:

1. Energy is never localized outside of matter.

2. The presence of energy in a body does not increase the inertia of the latter.

3. The presence of kinetic inertia in particular, that is, the state of motion of a body, does not increase its mass.

4. Absorption or emission of energy by radiation neither increases nor diminishes the mass of a body.

5. The mass of bodies being indestructible, the principle of the conservation of mass is distinct from the principle of the conservation of energy.

3. THE DIFFICULTIES OF THE DUALISTIC THEORY AND THE CHECK GIVEN TO THE MONISTIC ATTEMPTS AT REDUCTION.

The fundamental dualism of matter and energy leads to insurmountable difficulties. How can immaterial forces act on inert bodies so as to move *mind moves matter*

them and verify the ancient formula *mens agitat molem*? How can luminous radiation if it is devoid of mass, exert a repulsion in one direction on the material source emitting it, and a propulsion on an opaque body absorbing it, as if it represented an amount of momentum; how, in other words, can it, if it is destitute of inertia, behave like a material projectile, which exerts a recoil pressure on the fire-arm which throws it and a ballistic action on the obstacle which it strikes? And if energy possesses some inertia, how can it accumulate like electric energy on a body to a considerable amount, without the mass of the latter being found to have grown in some manner? In the presence of such an antinomy our understanding feels outraged and our imagination declares itself inadequate.

Furthermore there are the attempts at reduction, by the monists, coming one after another and seeking to resolve this initial contradiction by reducing one of the two terms of the antithesis to a mere semblance of the other. But Stallo, Hannequin, Duhem, and Meyerson[1] have shown that in these attempts the difficulty is simply shifted.

The first attempt, in time, is that of the atomists, whose fate, from Leucippus and Democritus to Huygens and the Bernouilli brothers, is especially noteworthy. It consists in reducing the supposedly

[1] Stallo, The concepts of modern physics. Hannequin, Essai critique sur l'hypothèse des atomes dans la science contemporaine.—Duhem, L'Évolution de la Mécanique.—Meyerson, Identité et Réalité.

occult notion of force to the clear and distinct one of mass; and for this purpose seeks to give account of all phenomena by the motion of elementary masses, indivisible and indeformable, impinging on one another. But if the atoms are rigid, transmission of motion through impact is impossible; if they are elastic, they are then deformable and composed of parts, which is contrary to the hypothesis and implies forces of cohesion and elasticity; this inevitable dilemma leads to shipwreck. Moreover Hertz,[1] in his posthumous mechanics, aims at explaining everything, not by atoms, but by articulated systems, by masses subject to firm bonds which must conform to a single law: every isolated system traverses with constant speed a trajectory of least curvature. What we take as real forces are fictitious forces of connection, due to the presence of bodies that we do not perceive, or fictitious forces of inertia, arising from a motion that we do not suspect. For example, anyone who, pulling a body tied to another by an invisible cord, saw the second move forward, would believe in a mutual attraction of the two bodies, while it would be the case of a force of connection produced by a hidden mass; and anyone who, not knowing the motion of rotation of a gyroscope, encountered an active resistance to an attempt to produce a deviation of its axis, would believe that a real couple tended to

[1] Hertz, Die Principien der Mechanik in neuem Zusammenhang dargestellt (Ges. W. Vol. III) Johann Ambrosius Barth, Leipzig 1894.

maintain this axis in an invariable direction, while it would be a case of a force of inertia produced by a hidden motion.

Thus "what we are accustomed," says Hertz, "to denote by the names of *force* and *energy* is nothing more than the action of mass or motion." As, moreover, nothing limits the motions and the hidden masses that may thus be introduced, it seems as if it were always possible to give an account of the behaviour of natural phenomena in this manner. But this very elasticity is in itself the cause of sterility. In spite of all the interest which is attached to attempts of this kind, such as those of Maxwell on electricity, of Lord Kelvin on the gyrostatic ether, and of Helmholtz on cyclical systems; the concept of Hertz has not led to a single positive result. Boltzmann[1] has shown that, however great the number of masses and hidden motions one imagines, and, however great, the consequent number of arbitrary variables to which one resorts, it is impossible to represent the simplest phenomenon in a satisfactory manner. Later, we shall find the reason for this lack of power to be the incompatibility of the form which the relativity principle imposes on physical laws, with the form of the equations of classical mechanics which govern the displacements of the elementary masses of the atomists and the articulated systems of Hertz.

The inverse reduction, of mass to force, has been

[1] Boltzmann, Anfrage die Hertz' sche Mechanik betreffend (Wiedemann's Annalen, Suppl. 1889).

the work of Boscovich, following Kant. The atoms
lose all material existence, and all spatial meaning;
they are nothing more than dynamical points, or
force centers. These forces are directed in the
straight line joining them, whence their name of
central forces. They are transformed in the nick
of time from repelling to attracting and their intensity depends on a certain power of the distance.
But if these force centers are mathematical points
in space, how can we imagine that a force can apply
itself to them in order to repel them or attach itself
to them in order to attract them? How can these
entities of thought resist motion, and exhibit inertia? "No arrangement of centers of force," declares Maxwell,[1] "however complicated, can account
for this fact; no part of this mass can be due to
the existence of the supposed centers of force." It
seems, moreover, that certain phenomena, such as
crystallization and permanent deformations, can
not be explained if one is limited to purely central
forces.

Lastly Lord Kelvin's[2] concept of a gyrostatic
non-dynamical ether, in supressing the rebellious
concepts of force and mass, has not thereby been
more successful. Reviving the ideas of Descartes,
Lord Kelvin contemplates a perfect fluid, homogeneous and incompressible, which fills all space. In
this fluid there exist eddy rings, which are eternal

[1] Maxwell, Theory of Heat, p. 86.
[2] Cf. Tait, Lectures on Some Recent Advances in Physical
Science.

and can neither be cut nor penetrated and which play the part of vortex atoms. Forces are due to pressure of the medium between these vortices. But, as Maxwell[1] remarks, one gets no sight of the invariable element which one would agree to consider as the mass of the atom. He contemplates therefore a pure motion in pure space without moving object, kinetic energy being half of the product of the square of a velocity by a zero mass, which is absurd.

Thus the mechanistic theory of the atomists, the kinematic theory of Hertz, the dynamical theory of Boscovich, and the non-dynamical theory of Lord Kelvin, have conclusively failed in their attempt to reduce either force to mass or mass to force. The efforts of this youthful dogmatism are followed by the prudent attitude of reserve of a school of physics that takes no interest in a mechanism explaining the phenomena, but describes simply, in its equations, the relations which connect the simultaneous variations of directly measurable physical quantities; a kind of physics in which there are many integral quantities, but no atoms.

Rankine and Ostwald have both remarked that these directly measurable physical quantities are always quantities of energy. "To establish such relations between measurable quantities that, some of these quantities being given, the others may be deduced, is the entire task of science. Hereafter there is no need of troubling ourselves about forces,

[1] Maxwell, Art. "Atom" in Encyclopedia Britannica.

the existence of which we cannot prove, exerted between atoms that we know nothing of, but we are concerned with quantities of energy put into play by the phenomenon under investigation . . . All the equations that connect one phenomenon with one or several of a different kind are necessarily equations between quantities of energy; there can be no others, for, apart from time and space, energy is the only quantity common to all kinds of phenomena."[1]

4. OSTWALD'S ENERGETICS.

From Ostwald's energetics, the genesis of which we have just traced, we may accept the following two propositions:

(1) Of the external reality we know only changes of energy and all physical phenomena may be described in terms of energy.

(2) In particular a body is only a complex of indissolubly associated energies, so that the concept of matter becomes included in the more general one of energy, and the principle of conservation of mass is absorbed by the more universal one of conservation of energy.

The first proposition is easy to justify. In the first place by psycho-physiology: The apparatus of our senses is set into action only by energy changes between it and the external world. The energy of the physico-chemical agents that im-

[1] Ostwald, Zur modernen Energetik, (Scientia, 1907, number 1).

presses it is transformed into nervous energy, which itself is transformed, in the cortical centers, into psychic energy *sui generis*. "If you receive a blow with a stick," says Ostwald[1] jocularly, "what do you feel, the stick or its energy?" The physical phenomena appear, on the other hand, as reducible to a condition of rest, transfer or transformation of energies of different forms: kinetic energy; potential energy of position; radiant energy; electric energy; magnetic energy; heat energy; and chemical energy. Every form of energy is the product of two factors, an intensity factor and a quantity factor, a variable of equilibrium and a variable of condition: kinetic energy is half the product of mass and the square of velocity; gravitational energy is the product of height and weight; volume energy, of pressure and volume; shape energy or elasticity, of force and displacement; electric energy, of potential and charge, or of elctromotive force and quantity; heat energy, of temperature and entropy; chemical energy of thermodynamic potential and the masses of the constituents.

The idea of a body becomes reduced to that of an energy complex: volume energy which causes a body to occupy a definite region of space; energy of motion which causes it to possess a definite capacity for kinetic energy or mass; gravitational energy which causes it to have weight. It is necessary to add to these three fundamental forms of

[1] W. Ostwald, Die Überwindung des wissenschaftlichen Materialismus. Zeitschr. f. phys. Chem., vol. 18, p. 305, 1895.

energy another which is essential for solid bodies, namely, energy of shape or elasticity, by virtue of which solids resist agents that tend to penetrate or deform them. Every body possesses, in addition to these energies, others in varying proportion: chemical, heat, electric, and magnetic energy. If we ask why energies of volume, motion, and gravity appear always associated in material systems, the answer is found to be in the very conditions of the external perception. If any one of these three energies were lacking in a body, the latter would escape our notice. A body not possessed of volume energy would not occupy space, it would be imperceptible; in the case of one not possessed of mass, an infinitely small impulse would give it an infinitely large speed but it would again be imperceptible; if not possessed of weight it would leave the earth and escape our observation. The objects that constitute the world of our sense experiences must therefore necessarily possess these three indissolubly associated forms of energy, the complex of which expresses all the positive contents of the idea of matter.

Two general principles govern the states of rest, transfer, and transformation of energy. They are the principle of conservation of energy and the principle of degradation of energy.

The first rests on the discovery of the mechanical equivalent of heat and its generalization for all forms of energy. It states that in an isolated system energy cannot be destroyed or created but can merely pass from one form into an equivalent

amount of another form. The second enunciates
that all forms of energy have a tendency to be converted into heat and all temperatures to become
equal, and may be also formulated thus: "In
a closed system the entropy always increases."
Boltzmann, by reducing the entropy to the logar-
ithm of the probability of a certain state, has shown
that this principle is a law of large numbers which
expresses, for very complex systems, the chance of
the most probable states being realized. In the
following we shall have to refer to the principle of
conservation of energy alone.

5. THE INSUFFICIENCY OF OSTWALD'S ENERGETICS AND THE EXPERIMENTAL DISCOVERY OF THE IN- ERTIA OF ENERGY.

In proposing his energetics as a method of
exposition, a gnosiological theory, Ostwald is unassailable. He is merely applying to physical
questions the positive method, as formulated by
Auguste Comte, who was gratified to see it applied
in the work of Fourier on heat; that is, he establishes
the equations that connect the simultaneous variations of measurable physical quantities, without pre-
occupying himself with the question whether they
can be reduced to each other qualitatively, according to their real nature. However, by excluding all
explanatory theories, Ostwald does not solve, but
eludes, the problem of knowing how an imponderable
agent by applying itself to a body with mass and
inertia can impart to it an acceleration or determine

a deformation in it.   He condemns us to an intellectual asceticism, a position which the human mind is not satisfied to take; and that too at the precise instant when the success of the hypothesis of electrolytic ions, advanced by Clausius and Svante Arrhenius, rehabilitated the more concrete theories, by making it possible, through the efforts of the latter, van't Hoff, and Ostwald himself, to group into a coherent body of doctrine the aggregate of electrolytic, osmotic and chemical properties of aqueous solutions.   Moreover, physicists have come by a different path to doubt the traditional antithesis between matter and energy and to establish by experiment the inertia of the latter.

The experiments of Faraday have demonstrated the localization of electric and magnetic energy outside of conductors and magnets, contrary to the first postulate of the dualistic doctrine.   The self-induction of electric currents in the conductors has revealed to them the existence of a veritable electromagnetic inertia.   The discovery of convection currents, predicted by Maxwell, realized by Rowland, has led them to foresee that, on the inertia proper of a charged particle in motion, there must be superimposed a supplementary inertia of electromagnetic origin, contrary to the second postulate of the dualistic doctrine.   This inertia, for velocities exceeding 30,000 km. per sec., varies with the velocity.   The cathode particles issuing from Crookes' tubes and the β rays from radium, issuing from the disintegration of the atom, have velocities of just this order of magnitude. Experimental study of the variations of the electromagnetic inertia as a function of the velocity becomes hereafter possible. Undertaken by Kaufmann and Bucherer, it shows the existence of grains of resinous (negative) electricity, or electrons, destitute of material support, the mass of which is solely of electromagnetic origin and varies as a function of velocity. Hence we have here a form of energy, electric energy, that is endowed with mass and structure, contrary to the fundamental principle of the dualistic doctrine. The relativity principle, which results from the failure of all attempts to demonstrate the absolute movement of a system by experience within this system, then intervenes to show that all bodies behave during translation as if their mass were solely electro-magnetic, that is, as if it were no more an invariable scalar quantity, but a tensor quantity, symmetrical and variable, contrary to the third postulate of the dualistic theory.

Absorption and emission of radiant energy by bodies and the pressure of radiation resulting therefrom, in accord with the theoretical views of Maxwell and Bartoli and the experimental results of Lebedew, lead, on the other hand, to endowing electromagnetic radiation with momentum, in order to safeguard the relativity principle and Newton's principle of action and reaction or its corollary, that of the conservation of momentum in an isolated system. But momentum implies moving mass. Radiant energy, therefore, has mass; and all other energy,

2

no matter what its form, being convertible into
energy of radiation, possesses a coefficient of inertia.
The mass of a material system changes through
absorption or radiation of energy, since, in the first
case, it is increased by all the inertia of the absorbed
radiation, and, in the second case, diminishes by
the amount appertaining to the radiation emitted.
From the formulæ it follows that, if we take the
velocity of light as unity, the inertia of a material
system is the measure of its internal energy. The
mass ceasing to be invariable, the principle of the
conservation of mass becomes merged in the principle of the conservation of energy, as previously conceived by Ostwald, and contrary to the last postulate of the dualistic theory.

Matter and energy are endowed equally with
inertia and, according to the experiments of Eötvös
on the proportionality of inert mass and gravitational mass, with weight in proportion thereto.
There still persists the opposition between matter,
characterized by its structure (the number and nature of the electrons, and possibly also the positive
remainders, of which it is composed) and the possibility of moving with velocities varying from zero
to that of light, and radiation which moves uniformly with the speed of light. This opposition
becomes weaker because of the *quantum* theory.
This theory shows that radiant energy is not uniformly propagated in a hypothetical medium, but is
emitted into space in the form of discrete units.

The metaphysical problem of the action of

an imponderable on a ponderable, of force on matter, disappears as a pseudo-problem, born of an artificial antithesis between matter and energy assumed at the outset. On account of the inertia of the latter, there is no more difficulty in understanding, for example, the pressure on the matter absorbing it, since, being endowed with momentum, light behaves strictly like a material projectile striking an obstacle.

Thus, as already stated, matter was erroneously opposed to energy, by denying to the latter everything that had been found to belong to the former. This led to a fundamental antinomy which gave rise to a pseudo-problem. The monistic attempts to reduce one of the terms of the antithesis to the other have merely served to shift the difficulty without relieving it, and Ostwald's energetics merely to escape the problem without solving it. The recent discoveries, however, by opening up a field of experiment of an extent beyond all expectation, such as the possibility of having at our disposal velocities approaching that of light, have led physicists to revise the primitive classification of natural agents into ponderables and imponderables. It has been discovered that this classification was not well founded, that the admitted antithesis between matter and energy was artificial, and that these two realities, in spite of differences which forbid their confusion, possess precisely the common characteristics that permit the explanation of their mutual action.

This result, essential for natural philosophy, is the end of a long line of theoretical reasoning and experimental verification, which unite conclusively to establish the following proposition: Energy is endowed with inertia. In order to understand the inner logic of this wonderful organic development, it must be followed without haste, step by step. The physicists have not found that royal road leading directly to the goal by which Ptolemy I asked Euclid to conduct him to the comprehension of the most transcendent scientific truths. They have not abandoned the old principles under pressure of the new facts until they had tried all possible ways of saving them. They have not risked one step forward without summoning an *advocatus diaboli* against themselves. They have been prudent in their innovations. Old mental habits, stabilized through long ages of scholasticism, and the narrow evidences offered by a traditional rationalism are not got rid of in a day; nor does one easily leave the beaten track of an imagination which derives satisfaction only from the contemplation of solids. In particular, the consideration of Maxwell's dielectric ether, which comes under the category of Hertz's hidden masses and movements, has for a long time, by misleading physicists into attempts at mechanical explanations, paralyzed efforts, of which the relativity principle finally had to show the futility. Leaving useless details aside, we shall follow in its essential curves the turnings of the winding road which has led the physicists of the new school to a conception of the universe, the beauty of which is a joy forever. To arrive there more easily we shall at the outset recall a certain number of ideas and principles which will be constantly used in the sequel.

CHAPTER II

MASS AND THE RELATIVITY PRINCIPLE

6. THE IDEA OF MASS; EINSTEIN'S EQUIVALENCE
PRINCIPLE AND NEWTON'S PRINCIPLE OF ACTION
AND REACTION.

In order to show how the idea of energy as endowed with mass and weight in proportion thereto
has been arrived at, it is necessary to define these
two ideas and to state two principles which are
closely associated with them.

What constitutes the individuality of a piece of
matter through all motions, divisions, recompositions, compressions, expansions, changes of state and
chemical combinations that it undergoes, is its mass,
which presents itself as an invariant of the group of
physico-chemical changes.[1] It is considered in
classical mechanics as an invariable scalar quantity
that characterizes every piece of matter and may be
defined in three different ways: (1) as a coefficient of
inertia; (2) as capacity for momentum; (3) as capacity for kinetic energy.

As coefficient of inertia, mass measures the resistance of a body to any action tending to modify
its state of motion. Newton assumes the propor-

[1] Cf. F. Enriques, Les Concepts fondamentaux de la Science,
p. 136–144.

tionality of the force acting on a body to the change
in velocity per unit time communicated to it, or the
acceleration; therefore the quotient of force by
acceleration defines the mass of a body. Moreover,
he assumes the principle of the independence of the
effects of a force, which leads to the view that the
mass of a moving body is independent of the velocity acquired: if a force acts on a body for a second,
starting from rest, and communicates to it the velocity v, the same force acting through another second
will communicate to it a second increment of velocity equal to the first, so that its velocity will become $2v$; if the same force continues to act through a
third second, the velocity will become $3v$ and so on.
Theoretically a moving body may be given a velocity as large as we please by the application of the
same force a sufficiently large number of times.
Lastly the mass of a body is independent of the
sense in which the force acts, whether it be parallel
to the direction of the velocity (tangential acceleration) or at right angles thereto (normal acceleration).

Mass may also be defined, as was done by Maupertuis, by starting with the idea of an impulse communicated to a body by a force during the time
element dt. This quantity is determined in magnitude and direction by the product $f \cdot dt$; and the
impulse of a body at the instant t is the geometric
sum of the elementary impulses that, starting from
rest, have been successively communicated to it by
the various forces exerted on it. In virtue of the
law of inertia, matter tends to conserve its state of

motion; the impulse is equal to the momentum of a body, or the product of its mass by its velocity taken in the same direction as the latter; hence the vector relation

$$g = mv.$$

Mass may therefore be defined, as *capacity for momentum*, by the impulse divided by the velocity according to the relation

$$m = \frac{g}{v}$$

Lastly, the kinetic energy of a body may be defined by the total work which had to be expended to bring the body from rest into its actual state of motion without deforming it for observers moving with it. This leads in rational mechanics to the relation

$$w = \tfrac{1}{2}mv^2.$$

Starting from this, mass can be defined, as the *capacity coefficient of kinetic energy*, by the quotient of twice the kinetic energy and the square of the velocity

$$m = \frac{2w}{v^2}$$

These three definitions are equivalent in classical mechanics and we shall have to refer to each of them in succession. But we shall see that in the new dynamics, the dynamics of the electron and of relativity, they cease to be identical for velocities above 30,000 km. per sec. All three lead to variable values, functions of the velocity, following three

different laws which make them assume an infinitely
large value for the limiting speed of light in a
vacuum. The following leads to a rejection of
Newton's principle of the independence of the
effects of a force. If a force acts for another
second its effect will be less than that produced
during the first; it will be still less during the third
and less in general as the velocity already acquired
by the body becomes greater, since the inertia
increases step by step, until it becomes infinite for
the speed of light. Lastly in the new dynamics the
acceleration communicated to a body by a force
depends on the angle made by the direction of this
force with that of the velocity. We must define a
tangential mass and a *transverse mass*, identical
with that defined by Maupertuis, which do not
behave symmetrically. Briefly, mass ceases to be
a constant scalar quantity and becomes a tensor
quantity, which is variable and unsymmetrical.[1]

The idea of mass designates two quite different
things: (1) the coefficient of inertia of a body which
measures its resistance to acceleration, or *inert
mass*; (2) the coefficient of attraction of a body on
another outside of it, according to Newton's law,
or *gravitational mass*. How do these two masses
behave with respect to each other? Newton supposed that there was a strict proportionality between the two coefficients; but that was not proved

[1] On the subject of symmetry of physical quantities see L.
Rougier, La symetrie des phénomènes physiques et le principe de
raison suffisante (Rev. de Mét. et Mor. mars 1917, p. 165–198).

experimentally before Eötvös[1] experiment with a torsion balance. On the surface of our globe all bodies are immersed in two superimposed fields of force: the field of gravity, which puts into play the gravitational mass of bodies, and the field of the centrifugal force, which puts into play their inert mass. The resultant of these two actions combined is the apparent weight of a body. If the two masses were different, the resultant, that is, the observed direction of vertical, would not be exactly the same for all bodies: there would not be a single vertical at one and the same point. That this is not so has been established by Eötvös by the aid of a torsion balance with an exactness which excludes relative divergencies of the order 10^{-7}, so that the exact coincidence of the two co-efficients may be taken for granted.

Einstein[2] has from this deduced the *principle of the equivalence of acceleration and gravitation;* the effects produced on any system by a gravitational field are not distinguishable from effects produced by a suitable state of acceleration of the same system when removed from the action of the field, or, what comes to the same thing, from those of a system referred to axes having an accelerated motion with respect to it. In order to learn the action of

[1] B. Eötvös, Math. u. naturw. Ber. aus Ungarn, Vol. VIII, 1890; Beibl, vol. XV, 1891, p. 688.

[2] Einstein, Ann. der Phys., 1911 p. 898 and seq.; 1912, p. 365 and seq.; 1914, p. 341 and seq.; Zeitschr. f. Math. u. Phys., 1913, p. 6 and seq.; Phys. Zeitschr., 1914, p. 176 and seq.; Archives des Sciences phys. et nat., 1914, p. 7–12.

gravity on physical phenomena it will suffice to study the effects of acceleration on them, that is, to examine the modifications that the physical phenomena undergo when passing from a reference system at rest, or in uniform translatory motion, to one accelerated with respect to the former. From this Einstein has made the deduction that if energy is inert it has weight in proportion. We can no longer imagine an agent like the ether having a certain density and yet absolutely imponderable, that is to say, devoid of weight. Everything that is endowed with mass must possess a certain coefficient of Newtonian attraction in proportion thereto.

The idea of mass is intimately connected with the Newtonian *principle of instantaneous equality of action and reaction*. Without attempting to fathom the meaning of this principle or seeking to find how it is connected with the definition of equality of masses,[1] we shall restrict ourselves to its enunciation. If a material point A acts on another B, the body B reacts on A, and these two actions are two instantaneously equal and oppositely directed forces. For example, the attraction exerted by the earth on the moon is at every instant equal to and oppositely directed to the attraction exerted by the moon on the earth. It follows from the principle of the equality of action and reaction that in the case of a closed system the momentum $g = mv$ remains constant. If a body through an internal action calls

[1] Cf. Mach, History of Mechanics, p. 210–216. Enriques, l. c., p. 144–149.

forth the appearance of a certain momentum, counted as positive, by acting on another body, it will
undergo a repulsion on the part of the latter and
consequently take a momentum equal to the first
and directed in the opposite sense so as to com-
pensate it. The center of gravity of the system
will remain at rest. The *principle of the conservation
of momentum* in a system which is subject to internal
actions only presents itself therefore as a corrollary
of the principle of action and reaction and would
cease to be exact at the same time as the former.

7. THE RELATIVITY PRINCIPLE.

The relativity principle originated from the negative result of all the attempts to show the absolute motion of a material system by experiments
within that system.

Classical mechanics assumed that all mechanical phenomena in a system in uniform rectilinear
motion are produced exactly as if the system were
at rest. An observer enclosed in Jules Verne's
bomb-shell could detect accelerations or rotations
that this shell undergoes by observing phenomena
taking place inside; he could in no way detect his
uniform translatory motion. This is no longer the
case if he has recourse to optical or electrical experiments. The electromagnetic theory introduces a
medium at rest, the ether, which transmits the
transverse waves of light with a definite speed, just
as air transmits sound waves. In the case of a
source of sound the relative motion of the source with

respect to the air can be measured without revealing an absolute motion of the source with respect to the ether, since air is dragged along by the motion of bodies. This is not so in the case of the ether. The latter, if it exists, is incapable of motion, so that one might hope to show, by electromagnetic or optical experiments within a system, the motion with respect to the ether of a source of light connected with the system.[1]

The earth possesses in its annual motion a translatory velocity varying constantly by amounts as high as 60 km. per sec. for the relative velocities corresponding to two diametrically opposite positions in its orbit, in January and July, for example. Let us consider the optical experiment actually performed by Michelson in 1881 and repeated by him and Morley in 1887 the outline of which is as follows. Let three points O, P, and P′ be taken at the vertices of an isosceles triangle with a right angle at O. OP corresponds to the direction of the translatory motion of the earth, OP′ is at right angles thereto; at O a glass plate is fixed inclined at 45°; at the points P and P′ mirrors are fastened normally to the two directions OP and OP′. Let us assume that a light ray comes from a lens on the line OP′ produced, strikes the glass plate O and there is divided into two rays: one will be reflected towards the mirror P,

[1] Cf. H. A. Lorentz, A. Einstein, H. Minkowski, Eine Sammlung von Abhandlungen, Teubner 1913. Laue, Das Relativitätsprincip, 1911. P. Langevin, Le temps, l'espace, et la causalité dans la Physique moderne (Bull. Soc. fr. de Philos. vol.. xii, 1912).

return to O and passing through the glass plate will
fall on the lens L; the other will pass through the
plate O, be reflected from the mirror P′ and be
superimposed in the lens on the first ray. If the
apparatus is at rest, the paths OPO and OP′O being
equal by hypothesis, the times t and t' required to
traverse them will also be equal, and the light disturbances brought by the two rays to the focus of
the lens will coincide in such a manner as to give a
maximum of light intensity at this point. The
aspect of the interference fringes will, by reason of
symmetry, remain the same for any rotation whatever of the apparatus, the locus of the points P corresponding to equal paths OP + PO being a sphere
with its center at O.

If the system has a uniform translatory motion v,
the locus of the points P corresponding to equal
paths traversed by the light, is no longer a sphere
about O but an ellipsoid of revolution flattened in
the direction of motion, the axis of which in a direction at right angles to the velocity is to that parallel
thereto in the proportion $1 : \sqrt{1 - \dfrac{v^2}{V^2}}$, v being the
velocity of the instrument, and V being that of
light. If the apparatus remains undeformed while
passing from rest to motion and if the instrument is
adjusted at the outset by experiment, so that the
paths OPO and OP′O are equal for a given initial
position, the aspect of the interference fringes will
be changed because a rotation through 90° will
interchange the directions OP and OP′ and this

change will reveal the absolute motion of the system.

For example, if the preceding experiment is made on the earth, in January, and the interchange of the directions OP and OP' does not bring about any change in the aspect of the interference fringes, we might interpret this negative result by saying that the velocity of the earth relative to the ether is relatively insignificant. Six months later it will have increased by 60 km. per sec. and the repetition of the experiment should bring about an appreciable change in the aspect of the fringes. Contrary to this anticipation the Michelson-Morley experiment gives a negative result, and so too do all optical and electromagnetic experiments undertaken by Lord Rayleigh and Brace, Trouton and Noble, Rankine and Trouton.

To account for this result, Fitz-Gerald and Lorentz assumed simultaneously that all linear dimensions of bodies contract in the direction of their translation in the ratio $\sqrt{1 - \dfrac{v^2}{V^2}}$, written more simply $\sqrt{1 - \beta^2}$. This contraction can not be ascertained by observers O, connected with the moving bodies, since all measuring instruments contract in the same proportion, but it can be by stationary observers O_0 who see it pass. Let us call the configuration of a moving body as measured by observers connected with it by means of measuring rods having the same translatory motion, its *geometric configuration;* and let us call its *kinematic configuration* the configuration of the same body measured by observers at rest with respect to it, who observe, at the same instant according to their clocks, during its passage, different points of its circumference, and determine the position of these different points by means of measuring rods graduated at rest; the kinematic configuration will not be identical with the geometric configuration, it will differ from it by a contraction in the ratio $\sqrt{1 - \beta^2}$.

In the reasoning that has just led us to anticipate a change in the aspect of the fringes as a result of the rotation of Michelson's apparatus, we have assumed the distances OP and OP' to be invariable during this rotation. According to Lorentz's hypothesis the distance OP', originally at right angles to the direction of motion, will, for observers O_0 at rest who see the instrument pass, contract during the rotation and become

$$OP = OP'\sqrt{1 - \beta^2};$$

while inversely the distance OP, originally parallel to this direction will expand in the inverse ratio and become

$$OP' = \frac{OP}{\sqrt{1 - \beta^2}},$$

so that the equality of the original times of travel t_1 and t_2 will carry with it the equality of the later times t_1 and t_2.

As a consequence of the Lorentz contraction spherical bodies become flattened ellipsoids in the ratio

$$\sqrt{1 - \beta^2}.$$

We shall have to use this result in the dynamics of the electron.

The negative result of experiments undertaken to discover the absolute motion of a system with respect to the ether has led physicists to generalize the relativity principle assumed in rational mechanics for mechanical phenomena:

Given different groups of observers, O_0 and O_1, one of which is in a state of uniform motion with respect to the other, the laws of nature will be exactly the same for the different groups of observers. This principle may also be stated thus: *the form of the equations expressing the laws of phenomena within a system does not change if they are referred successively to a reference system at rest and then to a second system having a uniform translatory motion with respect to the former.*

In order to see the meaning of the relativity principle thus extended to all natural phenomena in the case of uniform translatory motion, it will be best to compare it with the principle of relativity used in geometry and mechanics.

In Euclidean geometry we assume that the shape and the dimensions of a figure do not depend on its absolute position in space: they remain unchanged for all displacements that it undergoes and this is what constitutes the relativity of space. We may express this by saying that if the same figure be referred successively to different systems of axes O_0 and O_1, the equations of transformation that permit us to pass from the first system to the second,

3

form a group,[1] called an *Euclidean group*, in which the form of the equations that translate the properties of geometrical figures is an invariant. Geometry is nothing but the study of the structure of the Euclidean group, which is identical with the group of orthogonal substitutions.

In rational mechanics we assume that the mechanical phenomena taking place in an isolated system do not depend on its state of rest or of uniform translatory motion: this constitutes the relativity of motion. We may express this by saying that if we successively refer the same mechanical system to a reference system O_0 at rest with respect to it, and then to a system O_1 in uniform translatory motion with respect to the former, the equations of transformation that permit us to pass from O_0 to O_1 form a group, called a *Galilean group*, in which the form of the equations of classical mechanics is an invariant. In the simplest case we would have for such a transformation of the space coordinates of the first system with the axes x, y, z, into those of the second x', y', z'

$$x = x' - vt$$
$$y = y',$$
$$z = z',$$

where v is the velocity of translation of O_0 with respect to O_1.

---

[1] Given an aggregate of transformations we say they form a group, if the product of any two of these transformations and the inverse transformation of any transformation still form part of this aggregate.

We have just seen that in physics all phenomena whatever taking place within a system are independent of its state of rest or of uniform translatory motion, and this constitutes the relativity of all natural phenomena. This may be expressed by saying that if we refer successively the same physical system to a reference system O_0 at rest with respect to it and then to a system O_1 in uniform translatory motion with respect to the first, the equations of transformation that permit us to pass from O_0 to O_1 form a group, called a Lorentz group. Lorentz's equations of electromagnetism are an invariant of this group. In the simplest case of a translatory motion of the reference system O_1 parallel to the x-axis, the transformation of space and time coordinates of the first system with the axes x, y, z, t into the second x', y', z', t', will be expressed by the four relations

$$x' = \frac{x - vt}{\sqrt{1 - \beta^2}}, y' = y, z' = z, t' = \frac{t - \frac{\beta^2 x}{v}}{\sqrt{1 - \beta^2}},$$

where $\beta = \frac{v}{V}$, and where v represents the relative velocity of O_1 with respect to O_0 and V the velocity of light, which is the same in all directions.

Minkowski has given a very elegant geometric interpretation of the principles of relativity in a four-dimensional space. He substitutes in place of the Galilean reference systems formed by three axes at right angles in the Euclidean three-dimensional space, much more general systems formed by four axes at right angles in four dimensional space, in which time plays the part of a fourth imaginary dimension. Because time and the three space coordinates are thus strictly assimilated, Lorentz's transformation assumes a strictly symmetrical form. In fact, if the four coordinates are

$$x_1 = x, \qquad x_2 = y, \qquad x_3 = z, \qquad x_4 = ict$$

where i denotes $\sqrt{-1}$ and c the velocity of light in a vacuum, Lorentz's transformation becomes

$$X_1 = A_{01} + A_{11}x_1 + A_{12}x_2 + A_{13}x_3 + A_{14}x_4$$
$$X_2 = A_{02} + A_{21}x_1 + A_{22}x_2 + A_{23}x_3 + A_{24}x_4$$
$$X_3 = A_{03} + A_{31}x_1 + A_{32}x_2 + A_{33}x_3 + A_{34}x_4$$
$$X_4 = A_{04} + A_{41}x_1 + A_{42}x_2 + A_{43}x_3 + A_{44}x_4$$

The axes being at right angles, the coefficients A_{pq} satisfy the known relations

$$\sum_p A_{pq}A_{pr} = 0, \qquad \text{when } q \neq r;$$

$$\sum_p A_{pq}A_{pq} = 1.$$

The Lorentz transformation corresponds, therefore, to a *displacement* of the reference system of four axes (translation and rotation in the four-dimensional space). It occurs spontaneously in nature whenever an electromagnetic system is moved by a translation as a whole with respect to immovable axes. Hence the relativity principle may be stated as follows: *physical laws retain the same form for all systems of rectangular axes in four-dimensional space*, or again, if we call these systems Lorentz systems: *systems having Lorentz axes are equivalent.*

It follows as a result of the structure of the Lorentz group that the concepts of space and time cease to be independent concepts; they are absorbed in the more general concept of the universe. The duration of phenomena varies according as it is measured by observers relatively at rest or in motion. For every portion of matter there exists a *local time*, or *time proper*, which is the interval of time, measured by observers connected with the system, between two successive events in it which for them coincide in space.

Simultaneity loses all absolute significance and becomes relative: two events in two different regions of space which are simultaneous for certain observers cease to be so for others in motion with respect to the former. Even the order of succession of two events, the distance between which in space exceeds the path traversed by light during their interval, may be reversed for certain groups of observers.

These new ideas of space and time lead to the erection of a new kinematics. In it the idea of a natural Euclidean solid no longer exists, since all bodies undergo the Lorentz contraction in the direction of their translatory motion. If we still desire to talk about solids, it is necessary to substitute for Euclid's geometry that due to Lobatschewsky: the natural solids undergoing the Lorentz contraction behave therefore like hyperbolic solids in a space of negative curvature, taking for the value of the space constant the velocity of light $V = 3 \times 10^{10}$ cm. per sec.[1]

A second consequence consists in the impossibility of observing objects moving with a velocity exceeding that of light. This follows from the Lorentz contraction: at this limiting velocity bodies would become infinitely flattened out and have an infinite inertia. This may be derived also from the constancy of the velocity of propagation of light in all directions for any observers whatever as follows from Michelson's experiment. It follows from this experiment that a luminous ray takes the same time for the round trip OPO when it is propagated in the direction of the translatory motion of the earth, so that its velocity is combined with the velocity of that motion, and when it is propagated normally or obliquely to that direction, a fact that is explained by the existence of local time. Therefore, Galileo's law for the composition of velocities ceases to be true: two velocities v and v' in the same sense and the same direction may not be added by vector addition according to the classical relation $v + v'$, but must be added according to the new formula $(v + v') / (1 + vv')$,[2] where the unit of velocity is taken as the velocity of light: $V = 1$. All instantaneous action-at-a-distance becomes thereby impossible.

It is just this impossibility of any instantaneous

[1] L. Rougier, De l'utilisation des géométries non-euclidiennes dans la physique de la relativité (L'Enseignement mathématique, 15 janvier 1914, p. 5–18).

[2] This expression is always less than 1. (Tr.)

action-at-a-distance that leads to momentum, and hence mass, being attributed to radiant energy, if it is desired to preserve Newton's principle of the simultaneous equality of action and reaction or the principle of conservation of the total momentum in the case of a closed system.

Classical mechanics is in no better accord with the relativity principle than Galilean kinematics. There is an incompatibility between the Galileo group that conserves without change the form of Lagrange's canonical equations, and the Lorentz group that conserves the form of Maxwell's equations of the electromagnetic field. In order to put them into accord the coefficient of Lorentz

$$\sqrt{1 - \beta^2}$$

must be introduced in all the equations of dynamics.

It follows from this that all forces of nature must behave, on passing from one reference system to another, as if they were of electromagnetic origin.

In what follows we shall retain the following consequences of the relativity principle:

1. The existence of absolute motion has no physical meaning.

2. All bodies contract in the sense of their translatory motion in the ratio $\sqrt{1 - \beta^2}$.

3. No velocity exceeding the limiting velocity of light in a vacuum can be observed; consequently, there is no instantaneous action-at-a-distance.

4. All forces behave as if they were of electromagnetic origin.

We shall retain one last consequence. The incompatibility of the Lorentz group and the Galileo group shows the futility of attempts starting from the equations of Lagrange and Hamilton to explain mechanically the phenomena of nature, and in particular electric and magnetic phenomena.

On the contrary, what must be done is to reduce Lagrange's equations of mechanics, equations that are correct as a first approximation for low velocities, to Maxwell's equations of electromagnetism; the problem is to explain mechanical phenomena on the basis of electric phenomena.

CHAPTER III

ELECTROMAGNETIC DYNAMICS

8. THE LOCALIZATION OF ENERGY OUTSIDE OF MATTER.

The considerations that have led physicists to the idea of the inertia of energy are derived from Faraday's and Mosotti's views on the localization of electric energy outside of conductors in the dielectric space surrounding them.

It is known that the electrification of a body determines the appearance of attracting and repelling forces in the surrounding space, under the influence of which oppositely electrified bodies approach each other and similarly electrified bodies repel each other in accordance with Coulomb's law. Before Faraday, the appearance of these forces was explained by the statement that there exist two imponderable fluids, vitreous (positive) electricity and resinous (negative) electricity, to be found in the interior of electrified conductors, the quantities of which constitute the charges of these conductors. These fluids have the property of acting instantaneously on each other at a distance and this explains the electrostatic actions. The region of space in which a small electrified body may be directly influenced by the presence of electric fluid,

41

distributed inside of a neighboring conductor, is
called the *electric field* of this conductor. The
field is determined at every point of space by
the magnitude, the direction and the sense[1] of the
attracting or repelling force that would be exerted
on an electrified body brought there.

It is possible to draw through each point of an
electric field a curve whose direction coincides
with that of the force capable of thus acting at
that point: an infinite set of lines called *lines of
electric force* is thus obtained. These lines have a
purely geometric significance: that is, the attracting
and repelling forces, the direction of which they
symbolize, do not exist unless an electrified body
is actually brought to one of the points through
which they pass. In this case only do they acquire
a physical significance and that merely at the
points considered. The intensity of the electric
field of a conductor at a point is defined in magnitude, sense and direction, by the magnitude, sense
and direction of the force that acts at this point on
the unit of electric mass conventionally chosen.
If we call f the force exerted on the unit mass of
electricity η, the intensity of the field is represented
by the relation

$$h = \frac{f}{\eta}$$

[1] A force acts in a certain line (*direction*) and in a certain *sense*
(*i.e.*, either forward or backward) along the line. In works
written in English the word direction usually includes both ideas.
(Tr.)

This idea, which was the current one before Faraday, rests on three postulates: (1) there exist imponderable fluids (the positive electric fluid, the negative electric fluid, the positive magnetic fluid, the negative magnetic fluid and the caloric fluid) the presence of which in a body neither increases nor decreases its mass; (2) these fluids are distributed in the interior of bodies; (3) instantaneous action at a distance is possible.

Faraday rejects these postulates. By means of celebrated experiments he establishes the fact that the electric field is zero in the interior of conductors, that is, that their charge does not manifest itself except at their surface and in the surrounding space. He rejects the idea of instantaneous action-at-a-distance and considers the lines of force as having a real physical significance; they correspond, according to him, to certain permanent modifications of a dielectric medium, called the ether, filling all space and existing between the conductors. The mechanical state of this medium around an electrified body is what determines the forces to which this body is subject; and the electric charge carried by a conductor is determined by the number of lines or tubes of force attached to it. Starting from the concept of forces propagated by contact from point to point with the velocity of light, Maxwell has accounted for the electrostatic actions, in conformity with Faraday's views and in accord with his experiments.

It is possible to get rid of the consideration of the ether, the existence of which is hypothetical, and, as we shall see, contradictory, and contemplate nothing but the only positively accessible reality: electrostatic energy localized outside of the conductors in the form of the field. The mechanical work of attraction or repulsion, done by the static forces appearing in an electric field, represents a certain expenditure of energy, that is, the change of a certain amount of potential energy into actual energy of motion. Thus potential energy localized in empty space around a conductor exists in conformity with Faraday's experiments. This electrostatic potential energy comes, by virtue of the principle of the conservation of energy, from the work expended in producing the state of electrification of the conductor. The energy per unit volume is proportional to the square of the intensity of the corresponding field. Let $d\tau$ be a volume element taken in this space and let h be the vector defining in magnitude, sense and direction the intensity of the field at a point A of this element; then the density of the energy localized in the volume $d\tau$ has the value

$$\frac{K_0 h^2}{8\pi}$$

where K_0 represents the specific inductive capacity of space, the numerical value of which depends on the choice of units. To find the energy localized in a portion of space of finite extent it is sufficient to calculate the integral

$$W_e = \int \frac{K_0 h^2}{8\pi}\, d\tau$$

in this space.

Thus, for the empty space surrounding a conductor, Faraday's experiments and the principle of the conservation of energy lead to the assumption of *the localisation of a certain amount of electric energy outside of matter*.

The same considerations may be applied to a magnetic field produced by a magnet or a current. Similarly the energy localized in a region of empty space is equal to the integral

$$W_m = \int \frac{\mu_0 H^2}{8\pi} \, d\tau,$$

where H denotes the vector defining the intensity of the field at a point A of the volume element $d\tau$ and μ_0 the permeability of empty space.

The same conclusions are reached by considering radiation freely propagated in empty space with the velocity of light.

This radiation, if taken at a sufficiently large distance from its source, may be regarded as formed by the superposition of an electric and a magnetic field, one at right angles to the other and both normal to the direction of propagation.

This electromagnetic field represents a certain amount of energy outside of any material substratum, the energy density of which in every unit volume that it occupies is equal to $\frac{K_0 h^2}{8\pi} + \frac{\mu_0 H^2}{8\pi}$, the electric energy being equal to the magnetic energy in every freely propagated plane wave, so as to give the relation

(1) $$K_0 h^2 = \mu_0 H^2.$$

The localisation of electromagnetic energy of free radiation in empty space follows from the principle of the conservation of energy, as. may be readily demonstrated. Let us consider a material source radiating energy in a definite direction and an obstacle absorbing it completely. At start there is a loss of energy of the source and on arrival a recovery of this energy by the obstacle. The velocity of propagation of radiation being finite because equal to that of light, we would not have conservation of energy at every instant between the time of emission and that of absorption, unless radiation represented a transfer of energy through empty space proportional to the energy radiated at the start and absorbed on arrival.

Thus the principle of the conservation of energy and Faraday's experiments lead us to assume the localisation of electric and magnetic energy in empty space, outside of matter, around charged conductors and natural or artificial magnets. Nevertheless electrostatic potential energy is, in the case of a charged conductor, still closely bound to matter, for the vector representation of the electrostatic field shows the lines or tubes of force to be attached to the surface of a conductor and to spread radially roundabout. In the case of an artificial magnet formed by the magnetic field produced by a current flowing in a metallic circuit, the magnetic lines of force do not abut on the circuit: they embrace it by returning on themselves; but they still proclaim its immediate neighborhood.

This is no longer so in the case of radiant energy, which manifests itself to our senses sometimes as light, sometimes as heat, sometimes in the shape of Hertzian waves. Radiation represents a localisation, in empty space, of electromagnetic energy that is independent of the immediate neighborhood of any matter, and is capable of being propagated indefinitely in a given direction without ever meeting bodies which absorb it. Hence the first postulate of the dualistic theory must be rejected, namely that energy can not be localized outside of matter.

9. ELECTROMAGNETIC INERTIA AND THE DYNAMICS OF AN ELECTRIFIED PARTICLE IN MOTION.

If the first postulate of the dualistic doctrine has been set at nought as a consequence of Faraday's experiments, its second postulate has, since 1881, been shaken as a consequence of the theoretical researches of J. J. Thomson.[1] This young physicist had the merit of being the first one to understand that an *electrified body possesses, because of the electrostatic energy of its charge, a supplementary inertia of electromagnetic origin.* This results from the self-induction of conduction currents; from the existence of convection currents as a consequence both of the laws of Maxwell and Hertz of the electromagnetic field and Rowland's experiments relating to an electrified body in motion; and lastly

[1] J. J. Thomson, Philosophical Magazine, 5 ser. vol. xi, 1887, p. 219.

from the identity, as regards magnetic actions, of conduction and convection currents.

When the intensity of a current changes, a counter-electromotive force is produced in the circuit and this, in accordance with Lenz's law, tends to oppose the change of the current. This is self-induction which is a veritable electric inertia: it opposes the change in intensity of a current, just as inertia of matter opposes a change in velocity of moving bodies. Just as a certain effort must be expended to put a body in motion, a certain work must be expended to establish a current; and again, just as a projected body tends to conserve its velocity, so does a current once established tend to maintain its intensity. If a current increases or decreases under the action of an external field, a counter-electromotive force is produced which opposes the change of the current, just as the inertia force of a moving body that is accelerated or retarded opposes the external actions that produce this change.

The existence of convection currents was foreseen theoretically by Maxwell and was verified by experiment by Rowland and his pupils.[1] It results from the laws established by Maxwell and Hertz as to the interdependence which, in time and space, connects the change of the two fields, the electric and magnetic. They may be summarized qualitatively as follows: *any change in one of the*

[1] Crémieu, État actuel de la convection électrique (Bulletin des séances de la Société française de Physique, 1902, 3 fasc., p. 155).

two fields at a point in space as time goes on, gives rise to the other field, the field thus created being distributed in lines of force which encircle the direction in which the first field changes. A change in the magnetic field in a given region of space produces an electric field, the lines of force of which encircle the direction in which the magnetic field changes: this is the phenomenon of static induction discovered by Faraday. If the space where this field is thus created happens to be occupied by a conductor, currents that encircle the direction in which the magnetic field changes are induced in it. Inversely, a change of the electric field in a given region of space produces a magnetic field, the lines of force of which encircle the direction in which the electric field changes; this is the *convection current* foreseen by Maxwell and brought to light by Rowland's celebrated experiment.

Let us consider, according to this law, a point A taken in the electrostatic field of an electrified particle that is being displaced carrying its charge along with it. The intensity at A changes because the particle approaches this point, passes it, and moves away. There is a variation of the electric field in a fixed region of space as time goes on; therefore a magnetic field is produced on account of the law of the convection current. Hence an electrified particle in motion behaves like an element of a voltaic current; to put it more exactly, the charge e that it carries with a velocity v is equivalent, as regards magnetic actions exerted by it, to an

4

element of voltaic current having the intensity i and the length dl, so that $idl = ev$. The name *convection current* is given to the current thus produced.

Let us now consider an electrified sphere of radius a having a surface charge e. At rest it is surrounded by radial lines of force, distributed symmetrically all around, which correspond to a localization of electrostatic energy equal to the work that had to be expended to produce the state of electrification of the sphere, and which has the value

$$(2) \qquad W_0 = \frac{e^2}{2K_0}\, a.$$

If a rectilinear uniform motion with a velocity v, small compared with that of light, be communicated to this sphere, it carries along its train of lines of force, radially and symmetrically distributed as when at rest. On account of the variation, as a function of the time, of the electric field at a point A of space, it produces a magnetic field distributed in circular lines of force encircling the direction of motion. The field at the point A in a direction perpendicular to the plane passing through A itself, the instantaneous position O of the particle, and the direction of the velocity has the value

$$(3) \qquad H = \frac{ev \sin \alpha}{r^2}$$

where α denotes the angle AOv and r the distance AO. The field thus produced, which varies inversely as the square of the distance, is determined by the instantaneous position of the sphere. The latter carries along in its motion the system of circular magnetic lines of force superimposed on the radial electric lines of force. This system of lines of force constitutes the electromagnetic field of the moving particle, moving with it like the wave system following a ship. It remains invariable so long as the velocity remains constant. The magnetic field thus superimposed on the electric field represents a localization of energy, equivalent to the work expended by external forces in communicating the velocity v to the charged sphere. This energy that continues bound to it in its motion would be recovered when it is arrested, in the form of work done against the retarding actions. An easy calculation based on the expression for the volume density of energy $\dfrac{\mu_0 H^2}{8\pi}$ shows that it is equal to

$$(4) \qquad W_m = \frac{\mu_0 e^2}{3a}\, v^2,$$

which expression may be put into the more suggestive form

$$(4') \qquad W_m = \tfrac{1}{2}\left(\frac{2\mu_0 e^2}{3a}\right) v^2.$$

This energy presents all the characteristics of kinetic energy $\tfrac{1}{2}mv^2$, including that of being proportional to the square of the velocity. *Everything takes place as if, contrary to the second postulate of the dualistic doctrine, the sphere possessed, on account of the potential energy of its charge, a supplementary inertia, an additional kinetic mass of electromagnetic origin, equal to*

$$(5) \qquad m_0 = \frac{2\mu_0 e^2}{3a}.$$

This is a necessary consequence of the identity of conduction and convection currents. The inertia of a charged particle in motion is due to the self-induction of the convection current that it produces; since electric self-induction has the same properties as the inertia of matter, it follows that this particle, on account of its electrification, has an increment of its capacity for kinetic and, consequently, electromagnetic energy which becomes super-imposed on its inertia proper.

Electromagnetic inertia due to the presence of a charge on a conductor in motion is proportional to the electrostatic potential energy that this charge represents and that the sphere carries along with itself. Any change of this charge or of the radius of the sphere, causing a change of the potential energy stored around it, must imply a correlated change of this inertia. The question might then be raised, what would happen if the sphere had a velocity approaching that of light and its motion remained always quasi-stationary, that is, not subject to any appreciable acceleration. This is the theoretical problem that has been conceived by Max Abraham.[1]

In order to solve it we must refer to Maxwell's law on the mutual dependence of the electric and

[1] Max Abraham, Annalen der Physik, vol. x, 1903, p. 105–179.

the magnetic fields. The magnetic field produced at a fixed point in space by the moving particle changes with the time, according as the particle approaches, passes and recedes, carrying along its train of lines of force. This change of the magnetic field must, in accordance with Maxwell's law, be accompanied by the production of an induced electric field which is superimposed on the electric field given by Coulomb's law and modifies its distribution. The analysis of the phenomenon shows that the electric lines of force tend to place themselves in a direction transverse to that of the motion, and reach this state completely when the velocity reaches that of light. The limiting distribution of the field thus realized represents an infinite energy and consequently an infinite inertia; so that infinite work, requiring an infinite time, would be necessary to bring an electrified particle to a permanent condition in which it would have the velocity of light. This velocity appears as a limit that no electrified body can reach. This is confirmed by the fact that the velocity of cathode particles reaches nine-tenths of that of light without succeeding in attaining it; this, moreover, leads to the kinematic law of the composition of velocities, demanded by the relativity principle.

The modification of the electric field in consequence of the induced field superimposed on it carries with it that of the force of inertia that the particle opposes to changes of velocity. The electromagnetic mass ceases to be proportional to the velocity with a constant coefficient; it varies with the velocity according to a certain function, in such a way as to become infinite for the limiting velocity of light, and this in a different manner for the three definitions of mass and according to the direction of the acceleration with respect to that of the motion. In the simplest case, that of an undeformable spherical body, two masses must be distinguished; the longitudinal mass, the force divided by the acceleration, if the latter is tangential to the motion; the transverse mass if the acceleration is normal to the motion. The first case corresponds to a change in magnitude of the velocity without a change of direction, the second to a change in direction without a change in magnitude. We may call the value of m_0, which for low velocities is the same for the different definitions of mass, *initial mass* m_0 and reserve the name *electromagnetic mass* m for the transverse mass. This mass m is identical with the Maupertuisian mass; it is the only one accessible to the measurements on the electric and magnetic deviations of the cathode particles in a Crookes' tube.

To find what function connects the change of mass with that of velocity, the shape of the particle and the distribution of its charge must be specified. Max Abraham considers the case of an undeformable sphere having a surface charge of uniform distribution: for the different definitions of mass he arrives at rather complicated formulæ, all of which assign the velocity of light as the upper limit of the velocity of electrified bodies. Lorentz,[1] however, in order to account for the negative result of the experiments to discover the absolute motion of the earth, was led to assume the contraction of all bodies in the direction of their translatory motion in the ratio $\sqrt{1-\beta^2}$ for observers who see them pass. A moving sphere becomes a flattened ellipsoid. Allowing for this deformation, which modifies the distribution of the electric lines of force at the surface of a charged sphere and, consequently, the intensity of the convection current produced by it and the laws of the variation of the self-induction of the current as function of the velocity, Lorentz has obtained a formula much simpler than that of Abraham for the Maupertuisian mass or the transverse mass, which becomes equal to

$$(6) \qquad m = \frac{m_0}{\sqrt{1-\beta^2}}.$$

By starting from this formula the laws of variation corresponding to the other definitions of mass are readily obtained.

The fortunate parallelism, which existed at that time between the progress of theoretical physics and the discoveries of experimental physics, revealed in the cathode particles of Crookes' tubes, later in the β rays issuing from the atomic disintegration of radioactive bodies, electrified bodies

[1] H. A. Lorentz, Versuch einer Theorie der elektrischen und optischen Erscheinungen in bewegten Körpern, Leyden, 1895.

projected with velocities close to that of light.
It became possible to test whether their inertia
varied with the velocity, according to the theoretical
anticipation of Max Abraham, or according to that
of Lorentz, and to distinguish between the formulæ
proposed by these two authors.

The experiments of Kaufmann showed that the
mass actually varies with the velocity according to
the formula of Abraham; but the more recent experiments of Bucherer and of Hupka, of a higher
degree of precision, showed that the change of the
electromagnetic mass of β corpuscles is better
represented by the formula of Lorentz. We shall
see that the relativity principle leads to an extension
of this formula to all kinds of moving bodies, whether electrified or not, so that they behave as if their
inertia was solely of electromagnetic origin. It
follows from this that, contrary to the third postulate of the dualistic theory, the *mass of bodies is
not an invariable scalar quantity, independent of
their state of motion or of rest. It is a quantity having
the symmetry of a tensor,*[1] *and dissymmetrical and
variable as a function of the velocity.*

[1] In modern applications of *vector analysis*, the branch of
mathematics dealing with the algebra and calculus of directed
quantities or *vectors*, the concept of *tensor* plays an important
part. A *symmetrical tensor* is an aggregate of 9 quantities,
which, when combined with the 3 components of a vector by 3
symmetrical relations, give the 3 components of a new vector. A
non-symmetrical tensor is more complicated. Earlier writers
attached a quite different meaning to the word tensor; it was
used by them to denote the mere magnitude of a single vector.
(Tr.)

CHAPTER IV

THE ELECTRONIC THEORY OF MATTER

10. LORENTZ'S SYNTHESIS.

The interest of Kaufmann's and Bucherer's experiments is, however, not limited to the foregoing result. By subjecting cathode particles to the action of an electric field and a magnetic field the comparison of the two deviations thus produced permits of finding the velocity of these particles and the ratio e/m of their charge to their mass. If their charge is held to be constant and to be the same, these experiments then reveal the variation, as a function of the velocity, of the total apparent mass of these particles, consisting of the sum of their material mass and their electromagnetic mass. Comparing, then, the law of the total mass, as derived from the preceding experiments with the law of variation of the electromagnetic mass, theoretically calculated, the ratio of the material mass to the total apparent mass may be deduced. The result obtained is remarkable: *the inertia of cathode particles is solely of electromagnetic origin.* These particles constitute elementary electric charges without material support, atoms of negative electricity called electrons. Thus we have here, contrary to the dualistic theory, a *form of energy, resinous*

57

(*negative*) *electricity, that appears to be endowed with inertia and corpuscular structure without any material substratum.*

This result was the starting-point of the *electronic theory of matter*.[1] It consists in a change of everything for its opposite, a reversal of the relation traditionally supposed to exist between matter and electricity. In place of seeking to explain electric phenomena mechanically, the mechanical phenomena are now to be explained electrically. Instead of seeking, as Maxwell did, to cast the equations of the electromagnetic field in the analytical mould made by Lagrange and Hamilton, which appears no longer valid except as a first approximation for low velocities and for quasi-stationary motions, there is substituted for the dynamics of the material point, conforming to Newton's equations, the dynamics of the electron, conforming to Lorentz's equations.[2]

Matter, the substratum of mechanical phenomena, is explained by starting from electricity. The elementary charges of electrolytic ions, the cathode particles issuing from Crookes' tubes, the β rays issuing from the atomic disintegration of radioactive bodies, the negative emissions constituting the Edison and the Hertz phenomena reveal a universal constituent of matter in the electron.

---

[1] Cf. Les idées modernes sur la constitution de la matière, Conférences faites en 1912 à la Société française de Physique. Paris, 1913.

[2] Cf. W. Wien, Archives néerlandaises, 2. ser. vol. V, p. 96.

The molecular structures constituting ponderable bodies are nothing but electronic architecture. Electrons are grains of resinous (negative) electricity, all identical with one another. They carry an individual charge of 4.774×10^{-10} electrostatic C. G. S. units, distributed on a circumference of 2 millionths of a $\mu\mu$ (2×10^{-12} millimeter) radius. They are endowed with an initial electromagnetic mass equal to 10^{-17} grams, that is, 1800 times smaller than that of a hydrogen atom. They play different parts in the production of phenomena, according to the position that they occupy in the atomic and molecular structures and according to the degrees of freedom that they enjoy because of this position. Accordingly there are distinguished: perfectly free electrons moving outside of matter, in the form of cathode rays and β rays; free electrons moving inside of matter in the intermolecular spaces with a chaotic motion that gives rise to heat radiation, on which may be superimposed a translatory motion as a whole that gives rise to electric current in conductors; electrons weakly bound to atoms revolving on their circumference and capable of being detached easily under the action of shock or an electromagnetic action, producing thereafter ionization of solutions and gases; electrons bound to atoms, moving in closed curves, and determining the spectral rays as well as the chemical valences, the sources of molecular union; lastly, the electrons that compose the central kernel of the atom, inaccessible to external physical agents, in a chaotic state of agitation, which do not become revealed except in the atomic disintegration of radioactive bodies.

On this view the atoms are formed by negative electrons, bound more or less to the kernel, and of positive remainders, the charge of which must be equivalent to the total charge of the negative electrons in order to maintain in the neutral state the cohesion of the structure by the mutual attraction of the electricities of opposite kind. Of the structure of these positive charges we know next to nothing. Those that we can capture in the canal rays of Goldstein and the α rays of radioactive bodies are atoms of hydrogen or helium, deprived of one or more negative electrons and having, on that account, become positive ions. It may be asked whether there is a vitreous (positive) electricity, having an independent existence and a corpuscular structure like the resinous (negative) electricity, so that there would be positive electrons side by side with the negative ones, as Jean Becquerel believes can be inferred from the study of absorption spectra at low temperatures; or whether positive electrification results from the subtraction of one or more negative electrons from the normal atomic structure corresponding to the neutral state. In any case the question of finding whether positive electricity has a material support, so that there would be negative electrons but only ions of positive charge, has lost much of its interest. The relativity principle leads, as Lorentz has shown, to the assumption that the inertia of these positive ions, if they can

not be decomposed into simpler units, must follow
the same laws of change as a function of the velocity as that of the negative electrons, that is, that
these ions must behave as if their mass were solely
of electromagnetic origin.

The electronic theory has explained, with increasing success—which has only reached its limit
when confronted by black radiation and the diminution of the specific heats at low temperatures—the
phenomena of static electricity, of the electr c
current, of induction and magnetism, of the emission, the propagation in different physical media,
and the absorption of different kinds of radiation;
those of radioactivity, of the ionization of gases and
liquids, of chemical valence, etc. It has led to
the explanation of new phenomena such as the phenomena of Edison and of Hertz, and above all the
Zeeman effect. It has led Lorentz[1] to a brilliant
synthesis carrying his name. Presented for the
first time in 1892, this synthesis has been put into
accord with the requirements of the relativity principle by its author. According to it there is no
matter, only electrons, positive and negative, in
an ocean of uniform ether; all forces are of electromagnetic origin or behave like such; and the
measurements effected in a moving system are relative to the dimensions of the instruments, to the
forces taken as comparison terms, and to local time.

[1] Cf. H. A. Lorentz, Sur la théorie des électrons (Les quantités
élementaires d'électricité, ions, électrons, corpuscules, Paris, 1905,
vol. i, p. 430–476).

In this conception matter, conduction currents and magnetism become mere modes of manifestation. The only constitutive principles are the ether and grains of vitreous (positive) and resinous (negative) electricity. We shall see how the English physicists have taken hold of this conception and pushed the reduction in the number of things still further: according to certain of them not only is the existence of matter denied but also that of electricity; and nothing but ether and empty space exists.

11. THE DEMATERIALIZATION OF MATTER.

The conception of the dematerialization of matter of the English physicists results from the success of the electronic theory, taken in connection with the concept of an ether endowed with mechanical properties, that is, inertia, elasticity and rigidity, in conformity with the ideas of Faraday and Maxwell.

It is the wave theory of light, stated at first by Leonardo da Vinci and Galileo, made more precise by Huygens following the discoveries of Grimaldi, and established finally by Fresnel following Foucault's experiment that has given physical verisimilitude to the idea of an ether. If light and, in general, any form of radiant energy is a wave phenomenon that is propagated with a finite velocity in the interstellar spaces, it is difficult to avoid the assumption that, because there is wave vibration and therefore motion, there is something that moves, a mobile medium that must fill all space. As Lord Salisbury picturesquely said in his Oxford presidential address to the British Association, the first and principal reason for the existence of the ether is to supply a subject to the verb to undulate.

The second object of the ether is to account for the localization of energy in dielectrics, surrounding conductors or magnets, in accordance with Faraday's experiments, and for its accumulation in the neighborhood of a moving electric charge, in accordance with the theoretical views of Abraham and Lorentz and the experimental confirmations of Kaufmann and Bucherer. This energy, considered as a mode of manifestation of a substance, requires a substratum of which it is merely an accident. Lastly, actions at a distance, such as gravitational attraction, electric or magnetic attraction and repulsion, and, without doubt, chemical affinity, between bodies separated in space are incapable of mental representation except as pressures or tensions exerted by an intermediate medium in which the bodies are immersed. To account for the phenomena of gravitation, of optics, of electricity and magnetism, physicists have thus been led to consider a series of media, the discrepancies of which Hannequin[1] found pleasure in exposing, but all of which have in the end become fused together in Maxwell's dielectric ether.

It was the consideration of the ether that per-

---

[1] Hannequin, Essai critique sur l'hypothèse des atomes dans la science contemporaine, p. 178–224.

mitted a marked reduction of the number of primary principles required for an explanation of things. At the beginning of the nineteenth century the existence of eight energetic imponderable agents outside of matter was commonly assumed: the electric fluids, positive, negative, and neutral; the magnetic fluids, south, north, and neutral; the caloric, and the luminous fluid. Gradually these agents have disappeared: heat has turned out to be the sensible manifestation of the internal agitation of the molecules of bodies. Inspired by Fresnel's ideas Maxwell tried to explain the electric, magnetic and optical phenomena that ceased to be attributed to particular fluids by starting from the mechanical properties of the ether. Electrostatic phenomena, for example, seemed to him to be explained by the deformations of the ether. Static charges of electrified bodies have merely a fictitious existence: they are the locus of the ends of the lines or tubes of force corresponding to the deformations of the ether. These deformations subject the conductors to pressures and tensions forcing them to approach each other or to move apart, the electrostatic energy being nothing but the potential energy of deformation of the ether. The electrokinetic phenomena are explained by disturbances of the ether. The electric current is not a phenomenon taking place in the conductor but a state of motion of the adjoining ether, the character, direction, and intensity of which are determined by the geometrical and physical properties of the conductor, the energy of the electric current corresponding to the kinetic energy of the surrounding ether. There exist only two realities, matter and ether, the former of which acts like a supernumerary; the ether, in which, in the form of static deformations, the electric energy is stored, and in which, in the form of disturbances, the radiant energy is propagated, appears more and more as the only active medium taking part in the production of the phenomena.

Wherein has the electronic theory strengthened or weakened this idea, which satisfies our taste for simplicity? It seems at first that, in addition to matter and ether, it restores one of the old agents, the negative electricity, no longer conceived, however, as a continuous fluid but as a substance endowed with corpuscular structure, not as an imponderable but as endowed with inertia. In fact, we have the phenomena of electrolysis, which show us a monovalent ion carrying a well defined quantity of electricity e, which is always the same; a bivalent ion carrying a quantity of electricity $2e$ and so on. Now this elementary charge, which appears to be indivisible like an atom, is precisely that which is found again, but without material support, in the negative electrons, issuing from the cathode disintegration in Crookes' tubes, or from the atomic disintegration in the β rays of radium, or the kinetic agitation in the negative emissions of heated or illuminated conductors (Edison's and Hertz's phenomena). These charges can no longer play the

5

part of pure fictions, by being placed geometrically at the ends of lines or tubes of force constituted of deformations of the ether: they have a physical existence, a structure and inertia of their own.   In these grains of electricity, and not in the mechanical states of the ether, the initial cause of electric and magnetic phenomena must be sought.   The electric current, for example, has not its seat in the ether but in the conductor; it consists of a mass motion of the free electrons of the metal, superimposed on their chaotic kinetic agitation, like the flow of a gas in a pipe: there is no longer a conduction current due to the ether, but merely a convection current due to electrified particles in motion.

But at the very time when the discoveries of contemporary physics seem of necessity to increase the number of primary principles, by a return to the old agents, through assuming grains of resinous (negative) electricity, their offspring, the electronic theory, works an inverse reduction by destroying the concept of matter, which from this point of view has merely a gross semblance of reality.   According to this theory, the molecular and atomic structures constituting bodies are reduced to electronic architecture, the electrons being positive and negative, or simply negative, and their inertia being entirely of electromagnetic origin and due to the self-induction of the convection currents that they produce by their displacement.   Matter, ceasing to correspond to a distinct reality, is resolved into grains of electricity and these, according to the English school, become subtilized into mere ether cavities.

Rearming themselves with the ideas of Faraday and Maxwell, the English physicists, and the most famous of them, J. J. Thomson, regard the inertia of the electron as due to the ether surrounding it. An electron at rest is a surface charge without material support, and may be considered as a cavity in the ether. This cavity is the center from which the electric lines of force constituting the electrostatic field of the electron diverge. The ether adheres to these lines of force, so that the electron can not move without displacing it. The inertia of the electron results from the inertia of the entrained ether, which alone opposes its motion. Knowing the volume and the mass of an electron, the density of the ether adhering to it may be calculated; it is found to be equal to about 2000 million times the density of lead. Matter ceases to exist, since, being composed of electrons, the sole reason we have to believe in it, namely its inertia, does not properly belong to it, but is borrowed from the ether. Electricity likewise disappears as a substance with an existence of its own, since the electrons are reduced to cavities in the ether. Two principles only survive, ether and empty space, of one of which the ancients would have said that it is the non-being and consequently that it does not exist. The world is merely a bubble of ether in the non-thing.

This theory of the disincarnation of matter, leading to a complete etherization, should, in spite

of its alluring aspect, be taken with considerable
caution. It is threatened with downfall on account
of the concept that forms its base, that of an ether
endowed with mechanical properties, the hypothetical existence of which seems, moreover, to be
contradictory.

To fulfill its office the ether must accumulate
the mutually exclusive properties of solids and
fluids. It must behave like an elastic solid, endowed with a rigidity surpassing that of steel in
order to transmit nearly instantaneously the transverse vibrations of light; it must behave like a
fluid with a density much less than that of the
lightest gas, in order not to retard the translatory
motion of the stars and not to rob them of their
atmosphere; but in the neighborhood of the electrons it must have a density far surpassing that
of lead. All this is incomprehensible and no one
can, under the plea of thinking in terms of the
ether, evade thinking according to the law of
contradiction.

If the ether exists, it is incapable of motion,
as is proved by the impossibility of reconciling
Fizeau's experiment with Hertz's hypothesis of a
complete entrainment of the ether by matter in motion, and the impossibility of reconciling the principle of action and reaction with Fresnel's and
Fizeau's hypothesis of a partial entrainment of the
ether. But, if the ether is incapable of motion, our
laboratories and our instruments are continuously
traversed by an ether current, the velocity of which

is equal to and opposite to that of the earth and varies as a function of it. Such a current would exert a considerable influence on electromagnetic and optical phenomena. Now the experiments undertaken to show the absolute motion of the earth with respect to the ether, with which a privileged set of reference axes might be connected, show that there is none. The hypothesis of an immovable ether is in its turn contradicted by the relativity principle.

But there is more than that. Neither Maxwell nor any one after him has succeeded in giving a clear and distinct mechanical representation of the deformations and disturbances of the ether that would produce the electric and magnetic phenomena. In attempting it he arrived at a conception so strange, that of a cellular ether formed of two substances one of which was impregnated with the other like a sponge soaked with water, that he did not make it play a part in his large Treatise on Electricity and Magnetism. Recently Witte[1] has shown, by a very complete analysis, that the properties of the electromagnetic field cannot be explained by the aid of classical mechanics with any supplementary hypothesis, if we assume the hypothesis of a continuous ether, and he rejects as improbable that of a discontinuous ether. To-day we see the underlying reason of this impossibility in the rela-

[1] Cf. Witte, Ueber den gegenwärtigen Stand der Frage nach einer mechanischen Erklärung der elektrischen Erscheinungen, Berlin, 1906.

tivity principle. The Lorentz group, for transformations of which the form of the equations of electromagnetism remains unaltered, excludes the Galileo group, for the transformations of which the form of the equations of classical dynamics is conserved. Mechanical explanations of electric phenomena are, therefore, definitely condemned.

Lastly, we shall see that there are reasons for believing that radiation is not propagated by means of a hypothetical medium, but projected into empty space in the form of discontinuous elementary quantities, energy atoms, called *quanta*. With this return to an emission theory the most convincing argument in favor of the ether, the consideration of which becomes superfluous, disappears. This is expressed by Einstein:[1] "The electric and magnetic fields that constitute light no longer appear as states of a hypothetical medium, but as individual realities, which the luminous sources send into space as in Newton's emission theory."

If we renounce the mechanical ether of Faraday, of Maxwell, of Lord Kelvin and of Sir O. Lodge, and consider nothing but an absolutely empty space, at every point of which an electric and a magnetic field may be superimposed, then the idea of the dematerialization of matter gives place to that of a materialization of energy. The only reality that is positively observable and empirically

[1] Einstein, Phys. Zeitschr, vol. x, 1909, p. 849. Cf. Campbell, Phil. Mag. 1910, p. 981.

demonstrable is the energy that, at any time, is localized in a region of space and corresponds to the two fields, electric and magnetic, which are at the moment superimposed there, its density per unit volume being proportional to the squares of these fields. Nothing prevents us from considering energy as a substance, endowed in itself with existence, without the aid of any sub-stratum whatever. It is therefore convenient, instead of reducing the inertia of the electron to the agitation of the ether, to attribute it to the energy accumulated around it, which forms its electromagnetic wave system. Moreover we need not picture the electron as a spherical cell of two millionths of a $\mu\mu$ with a surface charge. It is more natural to suppose either that the density of this charge gradually diminishes from the center of the sphere outwards, so that the electron does not possess a definite surface, or that there is a uniform distribution of the electric density in the inside of the volume of the sphere, which implies a definite contour: in the latter case the same formulæ are obtained as in the case of a surface distribution, with the single difference that all quantities that play a part in the equations are multiplied by the factor $\frac{6}{5}$.

To sum up, a moving electron appears to us as a circumscribed region of space, where there is to be found accumulated electromagnetic energy in the form of fields, the intensity and distribution of which vary as a function of the velocity. The mass of an electron can not be attributed to a material support that does not exist, nor to the surrounding ether, the existence of which is hypothetical, but it is due to its own energy, which constitutes its only substantial reality. But if energy is inert it is endowed with mass, consequently with weight in proportion thereto, and, possessing structure, it becomes a materialized body, while matter is refined away.

If now, in conformity with Lorentz's theory, the molecular structures that constitute bodies are reducible to assemblages of electrons, and if the electron is inert only because of the energy which it possesses, material systems are themselves endowed with mass only in proportion to the energy that they contain. Mass becomes a quantity that measures their internal energy, and, as foreseen by Ostwald, the concept of matter is subsumed under the more general one of energy.

CHAPTER V

THE INERTIA OF ENERGY

12. THE MATERIALIZATION OF ENERGY.

The conclusion stated at the end of the last chapter appeared as almost an immediate consequence of a paper by Henri Poincaré[1] on the longitudinal contraction of an electron in motion. Taking the older point of view, that of the partisans of the mechanical ether, though the point of view thus taken matters little in this connection, he showed that this contraction is exactly that demanded for the maintenance of its equilibrium on the assumption that the surface charge carried by the moving electron, the elements of which tend to separate because of their natural repulsion, is maintained by a constant pressure of the ether. Assuming this uniform pressure, the equilibrium configuration of an electron at rest is that which makes the potential energy of the actions superimposed on it—electrostatic repulsions and Poincaré pressure—a minimum. If its shape is that of a sphere of radius a the total potential energy of an electron at rest in equilibrium is given by

$$(7) \qquad E_0 = \frac{e^2}{2K_0 a} + \frac{e^2}{6K_0 a} = \frac{2e^2}{3K_0 a}.$$

[1] H. Poincaré, La théorie de Lorentz et le principe de réaction (Archives néerlandaises, 1900, p. 252–278).

73

A comparison of this expression with that of the initial electromagnetic mass m_0 (5) leads at once to the relation

(8) $$m_0 = K_0\mu_0\, E_0.$$

According to the relation between the two coefficients K_0 and μ_0 deduced by Maxwell from the comparison of the two systems of C. G. S. units, the electrostatic and electromagnetic, namely,

$$V = \frac{1}{\sqrt{K_0\mu_0}}$$

where V represents the velocity of light, the expression (8) becomes

(9) $$m_0 = \frac{E_0}{V^2}.$$

The initial electromagnetic mass of an electron at rest is equal to its total potential energy divided by the square of the velocity of light. As the potential energy represents the only energy that an electron at rest can have, it is seen immediately that, the velocity of light being taken as unity, *the mass of an electron at rest is equal to its total energy, of which it may serve as a measure.*

This result may be generalized. If the electronic theory of matter were confirmed, it would be applicable to all material systems at rest, since they could be reduced to assemblages of electrons without material support. However it may be concerning this last point and postponing the question of the structure of the positive centers, whether material or not, Lorentz has shown that the rela-

tivity principle requires: (1) the longitudinal contraction of all bodies in the direction of their
translation in the ratio $\sqrt{1 - \beta^2}$; (2) that the change
of mass with the velocity, $m = \dfrac{m_0}{\sqrt{1 - \beta^2}}$, established
first of all for electromagnetic inertia, applies in
general to all bodies, as if their mass, like the forces
of elasticity and cohesion, were purely of electromagnetic origin. It follows therefrom that the
relation (9) must be applied to all material systems
at rest as if the electronic theory were correct.

This relation may be extended to all bodies
and material systems in motion. The same relation that exists between the masses m_0 and m of
the same body observed by observers O_0 at rest
with respect to it and by observers O_1 in motion,
must exist, by virtue of the relativity principle,
between the energies E_0 and E of the same system
observed simultaneously by O_0 and O_1, so that we
have

$$(10) \qquad E = \frac{E_0}{\sqrt{1 - \beta^2}}.$$

By virtue of (6) and (10), the relation (9) leads
to the generalization already stated, namely,

$$(11) \qquad M = \frac{E}{V^2}$$

The quantity $\sqrt{1 - \beta^2}$ being always less than one,
it follows from (10) that the energy of a body set
into motion without deformation is, for observers
connected with it, greater than when it is at rest.

The difference represents, by definition, the kinetic energy; and the two terms $\dfrac{E_0}{\sqrt{1 - \beta^2}}$ and m_0V^2, of which it is the difference, are nothing but two measures of the total energy of the same body, made successively by observers in motion and at rest.

To sum up, the existence of convection currents, foreseen theoretically by Maxwell and realized experimentally by Rowland, led J. J. Thomson, Max Abraham, and Lorentz to study theoretically the laws of variation of the inertia of an electrified particle in motion. The discovery of cathode particles permitted Kaufmann and Bucherer to verify these theoretical anticipations and to reveal the existence of grains of resinous (negative) electricity or electrons, the inertia of which is solely of electromagnetic origin and obeys the preceding theoretical laws. Starting from this result Poincaré has shown that the initial mass of an electron in equilibrium and at rest is equal to its total potential energy divided by the square of the velocity of light. The principle of relativity requires, then, that this result be extended to all systems of bodies at rest and in motion, so that the formula $M = E/V^2$, thereafter fundamental, shall hold for them. Taking the velocity of light as unity, this formula states that *the mass of a body is equal to its total energy, of which it may serve as a measure, and consequently energy is inert.*

The decisive arguments in favor of the inertia of energy are, however, drawn from considerations of another kind. They result from the necessity of reconciling Maxwell's pressure of radiation with the principle of relativity and with that of action and reaction or of the conservation of momentum in a closed system.[1]

Maxwell, by starting from the electromagnetic theory of light, and Bartoli, by starting from the principles of thermodynamics, foresaw theoretically, and Lebedew verified experimentally, that all radiation exerts a backward pressure on the source that emits it in a single direction and a forward pressure on an obstacle that absorbs it: this is what is called the *pressure of radiation*. The inertia of energy is a consequence of the existence of this pressure, and brings up the metaphysical problem of the action of an imponderable on a ponderable in a particularly acute form.

Let us consider first a material system the motions of which are due solely to internal actions, such as a firearm and its projectile. When the shot takes place the gun undergoes a recoil, that is to say, takes up a certain momentum which, counted negatively, represents a loss; the projectile is shot forward and acquires a momentum which, counted positively, is equal to that lost by the gun. The conservation of the total momentum of the system holds at every instant, and consequently there is no motion of the center of gravity of the system. The conservation

[1] Cf. A. Einstein, Ann. der Phys. vol. xviii, 1905, p. 639. P. Langevin, L'inertie de l'énergie et ses conséquences (Journal de Physique, juillet 1913, p. 553 et seq).

of momentum is but a natural consequence of the instantaneous equality of action and reaction. The gun recoils because, while it acts on the projectile, the latter reacts in turn equally on it.

Let us now consider a material source that radiates unsymmetrically in a single direction, such as a lamp provided with a reflector or a Hertzian exciter at the center of a parabolic mirror. At the instant of the emission the source recoils as a consequence of the pressure of radiation: there is a loss of momentum. If the radiation encounters an obstacle that absorbs it, it will communicate to it an impulse, that is, a momentum equal to that lost at departure by the source. The action experienced by the obstacle will be equal in magnitude to the reaction undergone by the source.

Does this mean that the principle of the conservation of momentum or of action and reaction is safeguarded? Certainly not, for it is not so at every instant from the start. A certain interval of propagation elapses between the time when the radiation is emitted and that when it is absorbed, during which the momentum lost by the source and the reaction that the latter undergoes remain uncompensated. This compensation would never take place if the radiation were propagated to infinity without encountering matter absorbing it. In this case there would be a definite loss of momentum and the center of gravity of the system formed by the source and the radiation would take an absolute motion, which is contrary to the relativity principle.

If this principle is to be safeguarded and at the same time those of the conservation of momentum and of action and reaction, the system source-radiation must be assimilated to the material system of the fire-arm and its projectile. The radiation must be treated as a material projectile, that is, it must be regarded as representing a certain momentum equal to that lost by the source, so that the reaction that the source undergoes is the natural effect of the action exerted on it; only in this case will the center of gravity of the system remain fixed and the relativity principle be satisfied. This is what Henri Poincaré has not hesitated to do. He was the first one to introduce the idea of electro-magnetic momentum of radiation to save Newton's principle of action and reaction, which is sacrificed in Lorentz's theory.

But momentum means, by definition, mass in motion, in virtue of the vectorial relation

$$g = mv.$$

The projectile carries away in its motion a part of the initial mass of the loaded fire-arm, which is diminished by just so much: that is why we have instantaneous equality of action and reaction, conservation of momentum and a stationary center of gravity of the system. Hence, if radiation is a vehicle of momentum, it must carry away with it a part of the initial mass of the radiating material source. It must be possible to assimilate literally the source-radiation system to that of the firearm-projectile, as also in the case of a radium atom during its transmutation when it spontaneously divides into a helium atom and a niton atom by a sudden explosion that projects the helium atom and the niton atom with equal momentum in opposite directions. After we have discussed the structure of radiant energy, which tends to make the radiation assume the characteristics of a corpuscular emission, the preceding assimilation will appear less bold. Hence if radiation carries away momentum it must, just like electric energy, possess electromagnetic mass apart from any material substratum.

Let us determine quantitatively the momentum and the Maupertuisian mass that must be thus attributed to radiant energy, to safeguard the principle of action and reaction and the relativity principle.

In the case of a plane wave, the electromagnetic momentum is in the direction of propagation and at right angles to the plane of the wave containing the two fields, the electric and the magnetic. By taking into account the fact that in this case the two fields are mutually perpendicular, it may be shown that the density g, per unit volume, of the momentum has the value

$$(12) \qquad g = \frac{K_0 h^2}{4\pi V}$$

where V represents the velocity of propagation of the plane wave, which is equal to that of light.

Since on the other hand, the density, per unit

volume, of the electromagnetic energy has, by virtue of (7), the value

$$(13) \qquad E = \frac{K_0 h^2}{8\pi} + \frac{\mu_0 H^2}{8\pi} = \frac{K_0 h^2}{4\pi}$$

the result for the density of momentum becomes

$$(14) \qquad g = \frac{E}{V}.$$

The electromagnetic momentum per unit volume is equal to the electromagnetic energy divided by the velocity of light.

To this momentum there corresponds a Maupertuisian mass by virtue of the vectorial relation $g = mv$.

The magnitude of this mass is, by definition, the momentum divided by the velocity, which, in this case, is that of light,

$$m = \frac{g}{V}.$$

If we replace g by its value E/V from the relation (14), we get

$$(15) \qquad M = \frac{E}{V^2}.$$

Free radiation has a mass equal to its energy divided by the square of the velocity of light. Since all forms of energy may be transformed into radiant energy, the preceding result may be generalized as follows: *every form of energy E possesses a certain coefficient of inertia determined by the preceding formula.* Contrary to the fundamental principle

6

of the dualistic theory *energy of any form whatever
is inert.*

The identity of the formulæ (11) and (15) shows
that the mass of a body and radiant energy are
equivalent quantities, capable of being converted
one into the other, as heat and mechanical work are.
If a body radiates energy, the radiation emitted
carries away a part $\Delta E / V^2$ of its initial mass, and
when an obstacle absorbs this radiation its previous
mass is increased by the entire Maupertuisian mass
$\Delta E / V^2$ of the absorbed radiation, whence:

*Any change ΔE of the internal energy of a material
system, due to emission or absorption of radiation, is accompanied by a proportional variation
of its mass in accordance with the relation:*

$$(16) \qquad\qquad \Delta M = \frac{\Delta E}{V^2}.$$

If we examine closely this enunciation of the
proposition as to the variation of the mass of bodies
as function of their velocity, it follows, contrary
to the third and fourth postulates of the dualistic
doctrine, that: *the mass of a body is not invariable;
it increases or it decreases according as the body absorbs or radiates energy, and as it is in motion or at
rest with respect to the system to which it is referred.*

The last postulate of the dualistic theory is
therefore necessarily unsound. In an isolated system, the different parts of which exchange energy,
the individual masses of the bodies present are not
conserved; it is only the total inertia of the system

(the constant sum of the variable inertia of the bodies and the variable inertia of the radiation) that is conserved, provided the system does not permit any exchange with the outside. The principle of the conservation of mass of material bodies is no longer valid by itself: it is replaced by the more general principle, which alone is valid, of the conservation of the total inertia of an isolated system.

How are we to evaluate this total inertia? An inspection of (9) and (10) shows that the total energy of a body, at rest as well as in motion, is equal to the product of its mass by V^2. If the velocity of light in a vacuum is taken as the fundamental unit, it follows that *the mass of a body is equal to its total energy*, a statement that translates the identity of the nature of mass and energy into a numerical equality; or, better still, *the mass of a body measures its internal energy.*

The total inertia of an isolated system, which neither loses nor receives energy, is, therefore, equal to its total energy: the internal energy of the bodies present, the kinetic energy of these bodies, the energy of free radiation. The principle of the conservation of the total inertia of a system goes back, therefore, to *the principle of the conservation of energy*, into which the principle of the conservation of mass is henceforth absorbed. In the new dynamics of relativity, there exist only two fundamental laws of invariance: that of the conservation of energy and that of the conservation of momentum. The first leads to the assumption of energy localized outside

of bodies; the second leads to the assumption
of the existence of electromagnetic momentum in
free radiation and consequently of the inertia of
energy. These two laws are not independent. If
we were to adopt Minkowki's terminology of the
"world" where the phenomena are referred to four
interchangeable axes, implying four homogeneous
coordinates, three of space and one of time, these
two laws would appear as two different aspects of
one single law, that of the conservation of the world
impulse.

Lastly, as a consequence of the exact proportionality between mass and weight shown by Eötvös,
from the inertia of energy follows its weight in
proportion thereto: a change of internal energy
would at the same time be accompanied by a change
of mass and a change of weight. Since it is easier
to measure the weight than the mass of a body, *the
weight of a body might serve to measure its internal
energy*.

13. THE EVALUATION OF THE INTERNAL ENERGY
OF BODIES AND THE VARIATIONS OF MASS.

From the inertia of energy the abandonment of
the dualistic theory and numerous special consequences result.

1. *Evaluation of the internal energy of bodies.*
Whenever a body loses heat or changes its dimensions under the influence of internal actions alone,
it performs thermal or mechanical work that
corresponds to a loss of some of its energy. The

internal energy of a body was formerly defined as the total work it could perform as a consequence of cooling without limits, or of extension or contraction without limits, according as the molecular forces are attractive or repulsive. There was no way of evaluating such a quantity of energy; changes only of internal energy could be measured by the work performed in starting from an initial condition.

On the other hand the formula $m_0 = E/V^2$ provides a very simple method of evaluating this energy of intermolecular and intra-atomic nature. It is equal to

$$E_0 = m_0 V^2$$

for a body at rest for those who observe it. This relation shows that a gram of matter at rest and at the temperature of the absolute zero corresponds to the presence of an internal energy equal to 9×10^{20} ergs, that is, an energy equivalent to the heat furnished by the combustion of 3×10^9 grams or 3,000,000 kilograms of anthracite.

Let us call the energy of a body thus evaluated for the temperature $T = 0$ and the state of rest *latent energy*. This enormous energy is nearly entirely of intra-atomic nature. In fact, at the absolute zero the degrees of freedom of the molecules are, as it were, anchylosed by the frost. On the other hand the physical molecular forces and the chemical atomic forces put in action a quantity of energy only very small compared with this enormous reserve of latent energy, as can easily be calculated. The changes of mass that result from the presence or absence of heat or kinetic energy or the presence of radiation within a body are practically imperceptible, except in the case of radioactive transformations, where the intra-atomic energy again comes into play.

2. *Change of mass with temperature.*—The same portion of matter, taken at two different temperatures, may pass from one to the other by emission or absorption of radiant heat. The change of mass resulting therefrom may be evaluated by dividing the heat exchanged with the outside by V^2. To calculate the order of magnitude of the effect anticipated, water may be taken, the heat capacity of which is especially large. A mass of water, having at 0° an inertia equal to 1 g., will have a larger inertia at 100°. The difference is obtained by dividing the heat absorbed, 100 gram degree calories or 4.18×10^9 ergs, by V^2, which in the same system of units is equal to 9×10^{20}, which gives about 5×10^{-12}, that is to say a quite imperceptible change.

Nevertheless this example shows that, from the theoretical point of view, the idea of mass must not any longer be confused with that of quantity of matter, as was done by Newton. Two masses of water of equal inertia, one taken at 100° and the other at 0°, do not contain the same quantity of matter, since they cease to be equal when reduced to the same temperature; to put it differently, two masses of water containing the same number of molecules do not have the same inertia unless they are taken at the same temperature, for then their energies are equal.

3. *Change of mass with velocity.*—The mass of a body depends on its state of rest or of motion with respect to given observers. In fact, when a body of initial mass m_0 acquires the velocity v, its mass increases on account of the kinetic energy acquired according to (6):

$$m = \frac{m_0}{\sqrt{1 - \beta^2}}$$

Its total energy E then becomes

$$E = mV^2 = \frac{m_0 V^2}{\sqrt{1 - \beta^2}} = m_0 V^2 \left(1 + \tfrac{1}{2}\frac{v^2}{V^2} + \frac{3v^4}{8V^4} + \cdot\right)$$

or

$$E = m_0 V^2 + \tfrac{1}{2} m_0 v^2 + \tfrac{3}{8} m_0 \frac{v^4}{V^2} + \ \cdot \ \cdot \ \cdot$$

From this formula it is seen that the quantity $\tfrac{1}{2} m_0 v^2$, called ordinarily the *kinetic energy of the moving body*, constitutes only a very small part of the energy corresponding to the passage from the reference system O_0, in which the body is at rest, to the system O_1, in which it is in motion. The quantity of latent energy $m_0 V^2$ remains hidden to our senses, which perceive only its exceedingly feeble variation. This formula shows, moreover, that kinetic energy loses its significance as a special form of energy, which is inevitable, for if one reduces inertia to energy one may not in turn reduce part of energy to inertia. This is why all forms of energy are functions of the velocity and grow with it.

4. *Change of mass with radiant energy.*—If the space comprised within a material enclosure is filled with black radiation corresponding to a given temperature, the mass M of the system will be, by virtue of the inertia of the radiation, larger than it would be without the latter: this excess mass is proportional to the total energy of the radiation.

5. *Change of mass in chemical reactions.*—We have already seen that the principle of the conservation of mass, if applied to the individual masses of the bodies present in a closed system, is generally erroneous. In particular, there is no conservation of the mass of the bodies in chemical reactions or in radioactive transmutations.

Chemical reactions being all exothermic or endothermic, it follows that, by virtue of the relation deduced from (16),

$$(17) \qquad \Delta E_0 = \Delta m_0 V^2$$

the sum of the masses of the elements combined does not remain constant.

Take, for example, the formation of water starting from its elements taken in the gaseous state. The combination of 2 grams of hydrogen with 16 grams of oxygen sets free 69000 gram-degree calories, equivalent to about 3×10^{12} ergs. We would not obtain 18 grams of water, because the heat liberated in the form of radiation carries with it a loss of mass equal to $\frac{1}{3} \times 10^{-8}$ grams, this being a difference of one-fifth of a billionth ($.2 \times 10^{-9}$) between the mass of the detonating gas and that of the water that it can form at the same temperature.

6. *Change of mass in radioactive transformations.*— The same will hold true of the transformations of radioactive bodies. The initial mass of one of these bodies and the total mass of its disintegration products at the end of a certain time will not be equivalent, the transformation being accompanied by radiation. It is known that one gram of metallic radium sets free 130 calories per hour while it is transformed into radium D and helium, through the successive forms of emanation, radium A, B, C. Taking into account the fact that the mean life of an atom of radium is about 2600 years, it may be computed that the complete transformation of one gram of radium into helium and radium D would liberate an energy equal to 1.1×10^{17} ergs. The emission of this energy would correspond to a difference between the initial mass of the radium and that of the radium D and the helium, equal, per gram, to

$$\Delta m_0 = \frac{1.1 \times 10^{17}}{9 \times 10^{20}} = 1.2 \times 10^{-4}$$

The disintegration of radium into helium and radium D represents merely one step of the transformations that start from uranium and end with helium and lead. The complete disintegration of a given quantity of uranium into helium and lead would represent a loss of mass exceeding one-tenthousandth of the original uranium. The fraction of the mass thus transformed into radiant energy is of an order of magnitude much greater than in the case of chemical reactions. It may be presumed that it comes from the latent energy of the uranium, that is, from its intra-atomic energy. If we could succeed in establishing exactly, to quantities of the order of magnitude 10^{-4}, the relation of the masses in the case of radioactive transformations, it would be possible to verify the identity of mass and energy.

To sum up, energy is inert and the mass of a body is equal to its internal energy which it serves to measure. This internal energy represents at the absolute zero an enormous accumulation of intra-atomic energy. According as a body acquires or gives up energy, its mass increases or diminishes. It is greater when the body is in motion than when it is at rest, greater when hot than when cold, when electrified than when discharged, it changes in chemical reactions and in a more perceptible way in radioactive transformations. The principle of the conservation of mass formulated by Lavoisier is true only as a first approximation: it becomes merged in that of the conservation of energy.

CHAPTER VI

THE WEIGHT OF ENERGY

14. THE WEIGHT OF ENERGY; ITS EXPERIMENTAL
    VERIFICATIONS.

The experiments of Eötvös show that if energy
is inert it must have weight in proportion thereto.
If this were not so, a certain quantity of uranium
and its disintegration products, helium and lead,
would have equal weights but different masses,
and consequently, would not be given the same
acceleration under the action of gravity. There
would have to exist at one and the same spot differences equal to at least one-tenthousandth in the
values corresponding to the acceleration of gravity,
and this seems to be capable of measurement.
Thus energy possesses not only an inert mass, but it
possesses also a ponderable mass $\mu = E/V^2$. In
accordance with what is true for inertia, a change in
internal energy is accompanied by a simultaneous
change of mass and weight. *A body is heavier when
in motion than when at rest, when hot than when cold,
when in a state of electrification than when neutral,
detonating gas than the water it produces, uranium
than its disintegration products.*

Langevin[1] sees an experimental proof of the

[1] P. Langevin, Journal de Physique, juillet 1913, p. 584.

91

inertia and of the weight of internal energy in the
departures from Prout's law. This law states that
the atomic weights are integral multiples of the
same quantity. While it is reasonably exact,
nevertheless the weights do present slight irregularities with respect to this law. These departures
would be caused by changes of internal energy
through emission or absorption of radiation ac-
companying the formation of atoms from the disintegration of primordial elements, as seen in
radioactivity or the inverse process of integration,
not as yet observed, with the formation of heavy
atoms. The sum of the weights of the atoms thus
formed would differ from that of the atoms transformed, by a quantity equal to the change in
energy divided by the square of the velocity of
light. These departures are such that the energy
put into play would be of the same order of magnitude as that actually observed in the course of
radioactive transformations. If, for example, the
atom of oxygen resulted from the condensation of
16 atoms of hydrogen or four atoms of helium, it
would be sufficient for an explanation of the atomic
weight 15.87 being less than 16, to assume that this
condensation is accompanied by a loss of energy
only five times greater than that set free during the
transformation of one atom of radium into radium
D. The interest of such an explanation of the
departures from Prout's law is to make the hypothesis of the unity of matter possible, that is, the hypothesis that all atoms are composed of one primordial

element or several such, and this can not be reconciled with these departures so long as the principle
of the conservation of mass in chemical reactions
is assumed.

Energy possessing a ponderable mass and the
inertia of a body being nothing but the inertia of
its internal energy, Newton's law expresses in
reality the *law of attraction of energy by energy*.
We shall see what follows from this for free radiation
and consequently for luminous radiation.

Radiation propagated freely in a vacuum represents, per unit volume, a certain energy density E
and a certain electromagnetic momentum equal
to E/V. It follows therefrom that it possesses
an inert mass, defined as the momentum divided
by the velocity, E/V^2. If every inert mass implies
the existence of a ponderable mass in proportion
thereto, a light ray will have weight; it will be
attracted by a mass situated in its neighborhood
by virtue of Newton's law. It will be deviated in a
gravitational field in proportion to the angle between the direction of the ray and that of the force
of attraction. Einstein[1] has calculated the magnitude of this deviation and, in 1911, arrived at the
formula

$$(18) \qquad \alpha = \frac{2KM}{V^2R},$$

where α is the deviation of the ray passing by a
spherical mass M (for example the mass of a star),

[1] Einstein, Ann. der Phys. vol. xxxv, 1911, p. 898.

K the constant of gravitation, R the distance from the center of the sphere to the ray. For a ray passing in the neighborhood of the surface of the Sun, α becomes equal to .83", which constitutes a quantity that can be measured, by observing, for example, the position of a star near the edge of the Sun at the instant of a total eclipse.

By virtue of Einstein's equivalence principle, luminous radiation in the interior of a system that undergoes an acceleration must behave like a projectile; instead of describing a straight line it will describe a parabola. Observers enclosed in Jules Verne's cannon ball, which would fall with acceleration, could therefore by the aid of optical experiments detect the state of acceleration of the system, without otherwise knowing whether they should attribute it to the presence of a gravitational field or to the state of acceleration of their cannon ball.

The equivalence between the inert mass and the ponderable mass carries with it the equivalence of the effects produced on physical phenomena by a field of gravitation and those due to a suitable state of acceleration of the reference system to which they are referred. It must follow therefrom that the potential of gravitation acts on the passage of time and the dimensions of bodies in the same manner as acceleration. Now it follows from the Lorentz group that a body is the more contracted in the direction of its translation for observers O_0 who see it pass and that the velocity of the phenomena that take place in it, measured by these same observers, is the more retarded, in proportion as its velocity is more accelerated relatively to them. In the same way, a body will be the more contracted and the march of the phenomena, of which it is the seat, the more retarded, in proportion as the potential of gravitation of the place where it is found is increased. Two equal chronometers placed at unequal distances from the Sun will go at different rates, and the more distant one will run ahead of the nearer. Now a chemically defined molecule that, by virtue of its oscillations, emits a given spectral light constitutes a chronometer of atomic dimensions. If, therefore two identical molecules at positions of different gravitational potential are observed with the aid of a spectroscope, one on the surface of the Sun and the other on the surface of the Earth, the oscillations of the second being more rapid than those of the first and consequently the frequency of the light that it emits being greater, it ought to be found that the ray emitted by the second is displaced in the spectrum, with respect to the ray emitted by the first, in the direction of the violet. Knowing the difference of the potential of gravitation at the surface of the Sun and at the surface of the Earth, it is easy to calculate that the difference in the wave lengths of the spectral rays ought to reach about $\frac{1}{100}$ Ångström unit, that is, one-millionth of a micron, which is a quantity accessible to experiment. It is remarkable that displacements of this order have actually been observed by Fabry and Buisson by comparing the Fraunhofer rays of the solar

spectrum with the corresponding rays of a terrestrial source.[1]

15. THE GENERALIZED PRINCIPLE OF RELATIVITY AND EINSTEIN'S THEORY OF GRAVITATION.

The relation $\mu = E/V^2$ shows that the laws of
the conservation of weight are the same as those
of the conservation of energy. Now the weight of
a body in a gravitational field changes when it is
displaced in the field: it increases when the body
is raised. To this increase in weight there must
be a corresponding increase in energy equivalent,
to the work expended against the weight in raising
the body. This change of energy carries with it,
in virtue of the formula $M = E/V^2$, either a change
of mass or a change of the velocity of light in the
gravitational field at the point considered. This
is the starting point of the new theories of gravitation according to the relativistic ideas.

G. Mie[2] has developed a theory in which he
makes the mass but not the velocity of light depend
on the potential of gravitation. He thus safeguards the postulate of the constancy of the velocity of light required by the relativity principle,
but he is obliged to renounce the equivalence between ponderable mass and inert mass established
by Eötvös. G. Nordström[3] has sought to maintain

[1] Cf. Freundlich, Phys. Zeitschr. vol. xv, 1914, p. 369.

[2] G. Mie, Ann. der Phys., vol. xc, 1913, p. 25; Phys. Zeitschr.,
1914, p. 115 and 169.

[3] G. Nordström, Phys. Zeitschr., 1912, p. 1126; 1914, p. 375,
604; Ann. der Phys., 1913, p. 533, 856; 1914, p. 1101.

the postulate of the constancy of the velocity of light and the proportionality between ponderable mass and inert mass over the widest range, but he is compelled to assume a change in the length of bodies and in the rate of phenomena, as a function of the potential of gravitation, which does not agree with that predicted by the Lorentz group. Einstein also has resigned himself to abandoning the constancy of the velocity of light in a field with varying potential and to assuming that it varies with position according to the formula

$$(19) \qquad V = V_0\left(1 + \frac{\phi}{V^2}\right),$$

where ϕ represents the magnitude of the Newtonian potential at the point considered. The Lorentz group is then applicable only to regions where the potential of gravitation is constant, or, what in virtue of the equivalence principle comes to the same thing, to systems in uniform translatory motion.

Einstein[1] has sought a group of transformations, having that of Lorentz as a special case, and such that the equations of the field of gravitation can be reduced to the form that those of a system without gravitation have, when referred to a reference system in a state of acceleration. The discovery of this group, for which he was indebted to the absolute differential calculus, created by

[1] Einstein, Die formalen Grundlagen der allgemeinen Relativitäts theorie (Ber. Berl. Ak., vol. XLI, 1914, p. 1030).

7

Christoffel and developed by Ricci and Levi-Civita, has enabled him to generalize the relativity principle by extending it to any arbitrary motion of the reference axes whatever, thus bridging the following serious epistemological gap, pointed out by Mach.

Psycho-physiology teaches us that our senses make known to us only relative states and changes of bodies, without ever revealing to us any absolute state or change. Thus our sight does not reveal to us the absolute shape and dimensions of a body, but simply the fact that a figure has such a shape and such a size relatively to another taken as comparison term, this purely relative shape and size not being altered in any dilatation or any continuous transformation whatever of the universe. In the same way, we do not perceive Newton's absolute space, to which we might refer all moving objects, but merely the relative state of rest or motion of bodies. Similarly our thermal sensibility does not teach us anything about the absolute temperature of bodies, but only about the changes in the states of affairs as regards heat losses between our skin and the surrounding medium. If, therefore, we wish to confine ourselves to the data of sense perception, we must not in natural philosophy speak of absolute velocity, acceleration or inertia.

Classical mechanics does not satisfy this requirement. It succeeds in preserving the relativity of velocities, but it concedes an absolute sense to the idea of acceleration and to that of inertia, considered as capacity of resistance to acceleration of a body. According to these principles the motion of two masses isolated in space and sufficiently close to be able to exert actions on each other, would be governed by Newton's law of attraction, independently of the system of fixed stars to which it is referred; for: "It would be," says Euler, "a very strange proposition and contrary to a host of other dogmas of metaphysics, to say that the fixed stars influence the inertia of bodies."[1] But, in fact, we perceive merely relative distances of bodies: hence we can only observe and define the relative velocities and accelerations of bodies, that is to say, the first and second derivatives of their distances. Consequently the inertia of a body can only be defined as its resistance to the relative accelerations which it experiences with respect to other bodies that do not participate in its state of motion. The inert mass of a body appears, therefore, as a relative quantity, which depends on the distribution of the masses about this body and on their state of rest or motion with respect to it: it will be greater, the greater the number of other masses in its neighborhood, which do not participate in its state of acceleration; it will disappear in the opposite case. A body is inert because it is surrounded by other bodies. Its inertia results from the mean action of all masses distributed in the universe, so that, contrary to Euler's assertion, the fixed stars do deter-

[1] Euler, Réflexions sur l'espace et le temps, p. 328.

mine in part the inertia and the motion of the Earth. The principle of inertia thus loses all absolute sense: it becomes a relative and statistical principle.

To put our physical conceptions into accord with the data of our sense perception, and to set natural philosophy free from metaphysical entities that encumber it, such as the ideas of absolute space and privileged axes, it is desirable to state the laws of physics in an intrinsic language, independent of any reference system, just as Euclid's geometry as compared with Descartes' is an intrinsic language free of the consideration of coordinates. For this purpose it does not suffice to make the form of the physical laws independent of the state of uniform translation of the coordinate axes; it must be made independent of any motion whatever of these axes; or, in Minkowski's geometrical language, it must be possible to refer them in four dimensional space to a system of oblique axes or a system of curvilinear coordinates, as well as to a rectangular system formed by rectangular coordinates.

Einstein's will be the glory of having succeeded in satisfying this last condition, by giving to the physical laws a universally invariant form, or, as we say nowadays, covariant for any change of coordinate axes. To accomplish this generalization of the relativity principle, Einstein started from the equivalence principle, deduced by him from Eötvös' experiments on the strict proportionality of the inert mass and the gravitational mass of bodies, which states the impossibility of distinguishing the effects of a field of gravitation from those of a field of motion. The effects produced by a field of gravitation can always be interpreted by a state of acceleration of a body removed from any field of force, and conversely. From this it follows that the existence of a field of gravitation in empty space is purely relative; it depends on the problem, insoluble by experiment, of finding whether the system from which it is observed is at rest or in accelerated motion; the real field for a system considered at rest will be fictitious for other systems in motion which the generalized relativity principle declares equivalent to the first. Einstein then shows that the form of the physical laws may be rendered independent of any system of privileged axes, provided the quantities characteristic of the field of gravitation are made to appear in the physical laws; or, more precisely, provided these laws be considered as relations between the quantities characteristic of material phenomena and the quantities characteristic of the field of gravitation, these quantities being tensors, and matter designating everything that is superimposed on the field of gravitation. These relations, being invariant for every transformation of the reference system, are *intrinsic equations, expressed with the aid of tensor equalities, from which every coordinate system has disappeared.*

This theory involves remarkable consequences.

It appears at first as a universal relativism expressed with the aid of an absolute calculus. The metrical properties of space, the kinematical and dynamical properties of mechanical systems, the physical properties of any region whatever of space vary according to the point of view of the observer. These properties depend, in fact, on the field of gravitation and the state of motion of the system from which they are observed.

The influence of gravitation is exerted on every physical process, on all matter including the electromagnetic and luminous field. Conversely, it has its origin in every region of the universe where the material tensor is different from zero. Since the material tensor corresponds to a reality, gravitation also involves a real element with which a fictitious field of gravitation, equivalent to an arbitrary motion in empty space, may always be combined.

Empty space is not the absolute and infinite void of Newton's followers: it is the pure field of gravitation on which no matter is superimposed. From the physical point of view there is no amorphous void, endowed with pure receptivity, in which material points could be imagined, attracting or repelling one another according to certain laws, like Newton's law, and thus communicating absolute accelerations to one another. We can not speak of empty space, except where there is a field of gravitation, not merely coexistent with this space, but veritably the creator of space, of its metrical properties, and, we may say, its extent. The world must be thought of no more as an assemblage of bodies lost in an infinite void, but as systems of bodies and of electromagnetic or luminous fields, superimposed on gravitational fields of finite dimensions. The absolute, void, amorphous, and infinite space of Newton vanishes like other *idola fori;* and Kant's antinomies advanced in regard to it are abolished as referring to a pseudo-problem.

To fathom the meaning of this universal relativism, it will be well, at this stage, to contrast the ideas of Einstein and those of Lorentz.

For Einstein there is no infinite void, no motionless ether, no uniform course of time, and consequently there are no privileged reference systems and clocks; there is no region of space enjoying absolute physical properties. The Lorentz contraction is not *true* in the sense that it corresponds to an absolute deformation, a body not having any shape except relatively to another. It is a reciprocal semblance of reality, arising from local time, from the fact that the clocks of observers connected with one body and those of observers in motion with respect to it, do not go at the same rate. The distinction between reality and semblance vanishes: there are only relative truths, science being unable to establish anything but comparisons, by ascertaining coincidences and by comparing colors (frequencies of radiating sources serving as clocks). The only absolute reality that it can attain consists of the laws of physical phenomena, expressed by intrinsic equations by the aid of tensor equalities, the form of which is independent of every system of space and time coordinates. If

God exists and it should please him to recount the history of the world in the extrinsic and artificial language of time and space, it would be merely by an arbitrary decree of his free will in a complete state of intellectual indifference that would make him choose, once for all, a system of space coordinates and a clock (that is, a radiating source connected with this system and consequently regarded as at rest). Having thus fixed arbitrarily the meaning of simultaneity and order of sequence of events, he could establish their universal chronology.

The ordinary distinction between semblance and reality would, on the contrary, exist for an infinite intelligence, if there be, as Lorentz is inclined to believe, an ether which is motionless but otherwise of a nature unknown to us, but it would be very different from the mechanical ether of Faraday, Lord Kelvin, and Sir Oliver Lodge. There would, therefore, be for an omniscient intelligence a privileged reference system and clock, whatever the system of axes, provided it be connected with the ether, and whatever the clock at rest, provided it be free from the influence of gravitation. This omniscient mind could with propriety speak of absolute motion, order of succession, and synchronism. The Lorentz contraction and the retardation of clocks would appear to him as physical phenomena, due to the connections between matter and ether, by which an action is exerted by the latter on the former. But this privileged set of three reference axes and this course of absolute time would be concealed from us forever as a consequence of the same action of the ether on the bodies, which would be precisely such as to prevent us from discovering their absolute motion. The observers O_0 would have no better ground for saying they were at rest and recording true time than the observers O_1. Any experimenter whatever could always explain what he observes by supposing either that he is at rest in the ether or that his laboratory is traversed by an ether current which produces the effect of shortening his instruments and retarding his clocks, or that there is no ether at all but it is the motion of a rod or a clock in his laboratory that produces the shortening of the one and the retardation of the other.

According to the pragmatist theory of truth the two concepts, that of Einstein and that of Lorentz, are equivalent, since they are equally in accord with phenomena.

On one hand, Einstein's concept, being in accord with the data of psycho-physiology, by respecting the conditions of knowledge imposed by our sense perceptions, by practising the scholastic adage *non sunt multiplicanda entia sine causa*, by relieving physics of metaphysical entities that encumber it, by delivering us from a swarm of pseudo-problems, is strictly positive, more economic than that of Lorentz, and unassailable in itself; on the other hand, that of Lorentz, by maintaining the ether, safeguards our old habits of thought, satisfies our craving for the absolute, sets our minds at rest, establishes a bond of union between the physics of yesterday and that of today, and falls into the category of explanatory theories, since the shortening of bodies, the retardation of clocks, the constancy of the velocity of light in a gravitational field of constant potential are explained by the connections of matter and ether. But the existence of the ether is forever rendered problematic by virtue of the relativity principle. Its nature is unknown to us, since the mechanical properties with which Faraday, Helmholtz, Lord Kelvin, Sir Oliver Lodge endowed it have already condemned it. It does not seem to represent necessarily the anticipation of future observations and experiments, like Pasteur's theory of microbes or the atomic theory; it is merely a method of exposition, a figurative hypothesis to uphold the spirit of abstraction, which, useful for the minds that Duhem called *broad* and *weak*, becomes superfluous, cumbersome and tiresome for the minds that he names *narrow* but *profound*.

Whatever the outcome may be, Einstein's results as to inertia, the weight of energy, and the relativity of phenomena seem permanent acquisitions. The relativity principle represents one of the norms of physical research which limits the field of our investigations and determines, in part, the form of the equations of physics.

16. ASTRONOMICAL VERIFICATION.

The philosophical importance presented by Einstein's theory of gravitation, based on the generalized relativity principle, should not lead us to forget the interest attached to its practical results, the most sensational of which, independently of those already indicated, is the calculation of the secular anomaly of the perihelion of Mercury.[1]

To account for it, Tisserand,[2] as is well known, proposed to replace Newton's law by a more general formula, similar to Weber's electrodynamical law. Of all the theories proposed since, none satisfies the mind like Einstein's theory of gravitation.

According to this theory, there exists no absolute invariant appertaining to a material point, that is, the exact equivalent of inert mass or of gravitational mass in classical mechanics. We have here to do with a tensorial theory in which the numbers G_{ab}, characteristic of the field of gravitation, and the numbers T_{ab}, characteristic of the field of inertia (ordinary matter and electromagnetic field), form a covariant tensor. It is possible to attach to these tensors two *relative* scalar invariants, G and T, which may, if desired, be considered by extension as measures of a gravitational mass and an inert mass.

Let us now consider what follows from this, taking account of the inertia of energy, in the case of the attraction of Mercury by the Sun.

In the study of the motion of a material point attracted by gravitational masses, a quite natural simplification consists in treating of this point as "a

[1] Einstein, Erklärung der Perihelbewegung des Merkur (Ber. Berl. Ak., vol. xlvii, 1915, p. 830).
[2] Tisserand, Traité de Mécanique céleste, vol. IV, p. 500.

test body" placed in a field of gravitation due to
other masses and not modifying perceptibly the
field of gravitation of these masses. Given two
bodies, an attracting body and an attracted body,
the inert mass of the attracting body (and the same
holds for the attracted body) is increased in proportion with the increase of the potential of gravitation due to other masses. Now the Sun, considered as an attracting mass occupying a sphere of
radius a, is surrounded with a field of gravitation, so
that it is immersed in gravitational energy. This
energy, always positive so far as concerns physical
reality, is negative in the sense that in order to
produce an attracting mass by condensation of elements coming from infinity, negative work must be
expended. It results from this that there is less
energy in the neighborhood of an attracting mass
than in the absence of all mass. If, therefore,
we attribute inertia to the energy of gravitation,
as to all the other forms of energy, it follows that,
for a point situated at a very great distance from the
Sun, the mass of the latter will be

$$M = M_0 - \Delta M_0,$$

ΔM_0 representing the correction term due to the inertia of the total field of gravitation. For a point
situated on the orbit of Mercury the effective solar
mass will be equal to

$$M + \Delta M,$$

if we denote by ΔM the supplementary mass due to
the inertia of gravitation outside of the orbit of
Mercury. It will thus be larger than the principal
mass M.

Taking account of these corrections, it is found that the elliptical motion is transformed into a pseudo-elliptic motion with a progressive advance of the perihelion. Applied to the case of the planet Mercury, the calculation indicates a secular motion of the perihelion equal to 43″. Since the observations give 45″, this aggreement must be regarded as the more remarkable, because it is obtained, without any supplementary hypothesis, by the application and simplification of the general equations furnished by the theory.

A hardly less remarkable confirmation of the theory of inertia of energy and of Einstein's theory of gravitation is obtained from the deviation of light rays in the neighborhood of the Sun. On account of both the Newtonian attraction exerted by the Sun on a light ray propagated in its neighborhood and of the curvature of space produced by the mass of the Sun according to Einstein's gravitational equations, the path of the ray should not be rectilinear but curved towards the center of attraction, with a total deviation given by the expression

$$\alpha = \frac{4KM}{V^2 R},$$

which is just twice the value given by equation (18), p. 93. In consequence of this it is possible to predict that a star seen near the Sun should suffer an outward deviation equal to 1.74″ and varying inversely as the distance from the center of the Sun for the more distant stars, as compared with the position that it occupies on the celestial sphere in the absence of the Sun, since the star is seen in the line of its ray, *i.e.*, in the direction of the tangent to the path of the ray. This deduction has been successfully verified by photographing the same region of the sky in the absence of the Sun and during a total solar eclipse. The negatives obtained by the two British expeditions undertaken for this purpose in the zone of the total solar eclipse of May 29, 1919, have given a mean deviation of 1.79″ + .03″. This agrees remarkably well with the value predicted by Einstein.

CHAPTER VII

˙ THE STRUCTURE OF ENERGY

17. THE SUCCESS OF THE ELECTRONIC THEORY;
THE EXPLANATION OF THE RELATIONS OF
MATTER AND RADIATION.

We have just seen that a body is inert and has
weight in proportion to the energy it contains,
so that the concept of matter is subsumed under
the more general one of energy, and that the principle of the conservation of mass becomes merged in
that of the conservation of energy. Energy, as
Ostwald would have it, becomes the only existing
reality, into which are absorbed the ether and the
numerous imponderable agents to which the physics
of the beginning of the nineteenth century was
partial. Nevertheless it appears that energy presents itself essentially in a double aspect: in the
form of resinous (negative) electricity endowed with
a corpuscular structure and in the form of free
radiation. In its first aspect it is made up of
grains of electricity, capable of moving with velocities ranging from 0 to V, its aggregates constituting
atomic and molecular structures, relatively stable
and with astonishing vacant spaces, and appearing
to our senses in the form of continuous bodies. In
its second aspect it appears as made up of transverse

waves, infinitely expansible and divisible, sweeping through all space with the uniform velocity of light. In the first case it takes the name of *matter;* in the second that of *radiant energy.*

The specific character of a portion of matter must no longer be sought in its mass and its weight, since radiant energy is likewise inert and endowed with weight, and mass is no longer an invariable scalar quantity, but takes the character of a tensorial quantity, which is unsymmetric and variable as a function of the velocity and the internal energy of bodies. That character must be sought in the number and nature of the primordial elements which constitute matter.

These elements, revealed in atomic changes beyond the domain of chemistry, are the electrons and the positive remainders, the latter having a structure as yet unknown and presented to us in the form of the positive kernels of helium atoms. These elements alone remain invariable throughout the changes that matter undergoes and can serve to define it.

What are the relations between matter thus characterized and radiation, and what is the mechanism of their exchanges of energy? It was on this point that the metaphysical pseudo-problem of the action of imponderable on ponderable arose, and it is to it that the electronic theory of matter claims to give a satisfactory answer.

Let us, for this purpose, consider again the case of a charged particle in quasi-stationary motion and let us see what happens when it undergoes an acceleration.

The wave system of the particle in quasi-stationary motion is formed by the system of radial electric lines of force and of circular magnetic lines of force that it carries with it. It may be, moreover, considered as the aggregate of the electromagnetic waves of velocity, emitted at different instants in its course, and centered on its former positions, which envelop one another, so that the electromagnetic field produced by the displacement of the particle is determined, not by the instantaneous state of the particle, but by all its former states. The energy of the field thus created is localized nearly entirely in the immediate neighborhood of the particle, because the intensity of the field varies inversely as the square of the distance, and rapidly approaches zero as the distance from the particle becomes greater. The waves of velocity which constitute the wave system of an electron do not correspond to any energy radiated to a great distance, since they vanish at infinity. They represent kinetic energy which accompanies the electron in its displacement, preserving for a constant velocity a fixed distribution around it. As space exerts no viscous action, no external influence is required to conserve this energy in motion, and the projected electron moves indefinitely with the same velocity in conformity with the principle of inertia, so long as no external cause modifies its state of motion and produces acceleration.

Let us examine what happens in the latter case. In the first place, the magnetic energy of the wave system, in conformity with the relation $\frac{e^2}{3a} v^2$ which is valid for small velocities, increases or diminishes by a quantity called *energy of change*, which corresponds to a reorganization of the lines of force of the wave system. In the second place the change of velocity determines the appearance of a spherical wave of acceleration, the radius of which increases with the velocity of light and which remains centered about the point where the electron was at the instant of the emission. This wave corresponds at every point to the passage of an electric field and of a magnetic field, situated in the plane tangent to the wave and perpendicular to each other in this plane. These fields represent a localization of electric and magnetic energy equal per unit volume, and this has the effect of giving to the wave of acceleration all the characteristics of free radiation.

The fields present in the wave of acceleration are superimposed on those of the velocity waves. Since the latter, varying inversely as the square of the distance, diminish much more rapidly than the former, which vary simply inversely as the distance, at a sufficient distance from the electron only the acceleration wave will exist. The energy thus radiated to infinity with the velocity of light, by the radiation which the electron emits during the time dt, is proportional to the square of the

8

charge and the acceleration according to the relation:

(20)
$$\frac{2}{3} \frac{e^2 v^2}{V} \, dt$$

where, in E. M. U., v denotes the acceleration and V the velocity of light.

This radiated energy represents the intrinsic energy of the acceleration wave. It is borrowed from the external actions which modify the velocity of the electron. But it represents only a small part of the energy which the acceleration wave carries at the start when leaving the center. The latter is the agency by which the electron reorganizes its wave system with the velocity of light, that is to say, by which the field of the electron receives the additional magnetic energy necessary for the increase of the kinetic energy associated with an increase of velocity, or restores the excess magnetic energy when its velocity diminishes. It is through the acceleration wave that the magnetic energy of change corresponding to an increase or a decrease of velocity is distributed to each part of the wave system or restored in the form of work, this magnetic energy being in the first case borrowed from external actions and in the second case restored in the form of work done against the retarding actions. The energy radiated from the acceleration wave represents as it were a necessary shrinkage, a tribute paid to the auxiliary wave for the service rendered.

The emission of radiation is always connected with acceleration of electrified particles. It is the difference in the conditions of electrons present in matter that enables us to explain the emission of different kinds of radiation and the phenomena traceable to them such as mutual induction and self-induction.

Let us consider first these last two phenomena. Conduction currents are reducible to convection currents: they correspond to the mass motion of the free electrons of the metallic or electrolytic conductors under the influence of a potential difference, the positive electrons tending to go in one direction, the negative in the other. If two wires are placed side by side, such as the two neighboring windings of a transformer, the passage of the inducing current corresponds to the circulation, in the first wire, of free electrons of which it is the conductor. The intensity of the current is proportional to the ensemble velocity of these electrons, and any variation of intensity will correspond to an acceleration experienced by the electrons. At the moment when the current varies, there will be an emission of acceleration waves by the electrons of the wire. The superposition of these waves at a point of the neighboring wire will give rise to.the appearance of an electric field parallel to the wire and in the opposite direction to the current if the intensity increases, in the same direction if it diminishes. In the case when this point is situated inside of the neighboring wire, the electric field thus created will determine a

current, corresponding to the existence of an electromotive force, the direction of which is given by
Lenz's law.

The phenomena of self-induction are explained
in the same manner. If the intensity of the current
increases, the acceleration waves emitted by the
electrons when their velocity varies will add themselves together in the circuit as well as outside, and
produce there an electric field, directed oppositely
to the current, by the agency of which the energy
necessary for the growth of the magnetic field
encircling the circuit is borrowed from the electromotive source that produces the current. If the
current diminishes, the accelerations of the positive
particles being opposed to the direction of the current, the radiated waves produce in the conductor
an electromotive field in the same direction, by the
agency of which the excess of energy of the magnetic
field, which encircles it, will be returned to the
circuit.

Conduction currents are at basis only convection currents; the explanation of their ˙ self-
induction must account for the inertia of an electrified particle in motion. This inertia is due to the
production of a magnetic field having its source
in the displacement of the particle, and to the fact
that the energy present in this field must vary with
the velocity by the agency of the acceleration
wave. The force of inertia that the particle
opposes to the change of velocity comes from
the action, on every element of its charge, of the

electric field present in the acceleration waves emitted by the other elements of this charge.

The character of the free radiation, which manifests itself in the form of Hertzian waves, of light, of Röntgen rays, and of thermal radiations, depends on the nature of the electrons that are accelerated and on the circumstances of their acceleration, sudden, continuous, or periodic.

A sudden acceleration takes place on the arrest by an obstacle of the cathode particles or the β particles, projected with velocities comprised between 20.000 km. and 290,000 km. per second. The radiation consists in a sudden pulsation, a kind of electromagnetic noise, emitted at the moment of the arrest of the particle, its thickness being equal to the product of the velocity of light and the duration of the shock, that is to say, it is of atomic dimensions. This extreme thinness explains the penetrating power as well as the absence of refraction of X-rays.

The acceleration is periodic in the case of the electrons that move in closed curves around a positive center inside the atoms. From it there results a continual emission of regular waves, the period of which is equal to the time of revolution of the electrons. This emission corresponds to light of a definite wave-length, like that which constitutes discontinuous spectra.

If the acceleration is due to the chaotic agitation of the free electrons of incandescent sources, there result from it radiations of every wave-length, which form the continuous spectra of incandescent bodies.

If the radiation is due to the accelerations which the free electrons of an opaque metallic enclosure undergo as a consequence of the thermal agitation of the molecules that strike them, it constitutes the heat radiation inside of this enclosure, called *black radiation*.

The presence of moving electrons in matter explains not merely the emission of radiation but also the laws of its propagation through material media and those of its absorption.

The absorption of radiation by matter is due to the fact that the electrons present in the matter are, under the action of the alternating fields of the incident electromagnetic waves, set into vibration in agreement with them. In consequence of the motions thus produced, they strike the neighboring material molecules, the kinetic agitation of which they increase at the expense of their vibratory energy. It is by this mechanism that the electromagnetic energy of the absorbed radiation is changed, through the agency of the shocks, into thermal energy, that is, into kinetic energy of chaotic agitation of the molecules of the absorbing body. It is because of this that a body which absorbs radiation is heated at the expense of the incident electromagnetic energy, and that its mass is increased by all the Maupertuisian mass of the absorbed radiation. In the case of selective absorption it is not the free electrons but the electrons in a regular periodic motion within the atoms which absorb, by a resonance phenomenon, the radiations of a period equal to that of their motions.

In conclusion, the electronic theory of matter appears to give an account of the mechanism of the relations existing between matter and radiation. By the success of this step, the electronic theory is correspondingly strengthened. It reinforces the view that at basis there exists only electromagnetic energy, which presents itself in two different aspects: narrowly circumscribed and concentrated in space in the form of elementary charges in motion, the velocity of which may vary from 0 to V; and in the form of radiation sweeping through all space with the rapidity of light—in the form of grains of electricity and in the form of transverse waves indefinitely expansible and divisible—stabilized in structures of varied and definite architecture, or free to lose itself by radiation to infinity.

The success of the electronic theory and of the mechanics of relativity reduces finally the primitive pluralism of ponderable matter and imponderable agents to the dualism of negative electricity (possibly also positive), endowed with a corpuscular structure and of electromagnetic radiation formed of continuous waves, both of them having inertia and weight.

18. CHECK TO THE ELECTRONIC THEORY: BLACK
RADIATION AND THE QUANTUM THEORY.

The electronic theory has nevertheless reached its limits: it has been revealed as incapable of giving an account of the law of the distribution of energy in the spectrum of a black body and of the diminution of the specific heats of solids at low temperatures, while the classical theory of radiation, due to Maxwell and Hertz, suffered shipwreck from certain phenomena of physical optics. To bring theory and experiment into accord new supplementary hypotheses had to be introduced. Are they going to modify the previously acquired results, to give support to a dualism between matter and radiation? Far from it, on the contrary they weaken it, for they amount to nothing less than endowing radiant energy with structure, after it has already been endowed with inertia and weight. In fact they finally lead to the view that radiation is not a system of waves infinitely expansible and divisible, propagated by a hypothetical medium, the ether, but a projection of matter in empty space with the velocity of light, in discrete units, emitted and absorbed by bodies in a discontinuous manner.[1]

In order to understand the point at issue, we must begin by defining what *black radiation* is.

Any body that is not at the absolute zero emits energy in the form of thermal radiation. When

[1] Cf. La théorie du rayonnement et les quanta (Rapports et discussions de la réunion de Bruxelles, publiés par P. Langevin et M. de Broglie, Paris, 1912).

unequally hot bodies are placed in an enclosure observation shows that they finally come to thermal equilibrium, all bodies in the enclosure reaching the same temperature. This equilibrium can have been attained only by exchange of radiation between the bodies. When it is attained the bodies do not radiate less, but each of them acquires by absorption as much as it expends in emission. This result has led Kirchhoff to state the following law on the subject of thermal radiation: the emissive power of a body for every kind of radiation and at any temperature is equal to its absorbing power. In the case of an ideal black body, which completely absorbs the radiations that strike it, the absorbing power is equal to unity. We are thus led to seek for the law of the distribution of radiant energy in the spectrum emitted by a black body at a given temperature.

In nature there exists no rigorously black body, no body that does not reflect or diffuse in part the radiation that it receives. Kirchhoff has enabled us to get over the difficulty by supplying the means of realizing a black body artificially. Let us consider an opaque enclosure and any radiation whatever that is propagated inside it: it will strike the wall a first time and will be partly absorbed; the remaining radiation, reflected or diffused, will again strike other parts of the wall and be there absorbed in the same proportion, so that the radiation which escapes absorption will tend rapidly towards zero. Such an enclosure possesses therefore an absorbing power equal to unity for any radiation: it realizes the black body.

Let us plunge this enclosure, after having evacuated it, into a bath at constant temperature, so as to maintain its walls at constant temperature. Experiment proves that the space inside this enclosure is isothermal, that is, that a thermometer placed at any point whatever inside of this enclosure, will finally indicate the same temperature. Any action on the thermometer placed in the vacuum must be exerted by radiation. In the region where the thermometer is placed, undulations arriving incessantly from different points of the enclosure become superimposed and form systems of stationary waves of definite frequency, adapted to the distances existing between two walls and realizing a permanent state of extremely rapid changes, the details of which are beyond our scale of time measurement. This is what is expressed by saying that the thermal equilibrium realized in the interior of the enclosure is a statistical equilibrium. This equilibrium is realized for every kind of radiation separately, and it is characterized by the amount of energy that is contained per unit volume in the space within the enclosure and by the distribution of this energy between the different wave lengths of the systems of stationary waves. It is precisely this energy density and this distribution of energy in the spectrum that has to be determined for black radiation.

This we succeed in doing by the following method. Experiment proves that the temperature indicated by the thermometer is independent of the nature, the shape and the dimensions of the walls of the enclosure. It follows therefrom that all the directions are equivalent and that each cubic centimeter of the enclosure contains the same density of radiant energy. If we consider in the enclosure a plane closed contour of one square centimeter surface, the amount of radiation that passes through this contour in one second will have a definite value, proportional to the density of the energy of radiation in equilibrium at this temperature. To study the composition of the radiation of the isothermal enclosure, which is identical with that of a black body and on that account is called black radiation, it will be sufficient to make a small opening in the enclosure, so as to collect the radiation that comes out of it and that we know to be identical with the radiation which in the isothermal enclosure traverses at every instant a section of equal contour. By this device we can deduce Stefan's law for the density of radiated energy: the density of the total energy radiated per unit time is proportional to the fourth power of the absolute temperature. By receiving the radiation in a dispersing apparatus it has been found, as regards the distribution of energy in the spectrum, that at a given temperature the energy radiated shows a maximum for a certain definite radiation on both sides of which it decreases rapidly. When the temperature rises the energy density of each simple radiation always increases, but the maximum of intensity is displaced towards the shorter wave-lengths, that is, for the visible part of the emitted spectrum, from the infra-red towards the ultra-violet. The curve of intensities becomes displaced along the spectrum, mounting more and more, at first slowly, then with an extreme rapidity the higher the temperature rises. Wien has discovered the law of this displacement as a function of the wave length for a given temperature, and this determines the distribution curve of the total energy of the spectrum of a black body: the wave length corresponding to the intensity maximum decreases inversely as the absolute temperature, while the intensity of the maximum increases in proportion to the fifth power of the absolute temperature.

It remained to express in a single empirical formula the distribution in the spectrum of the intensities of the black radiation for all the temperatures observed: this is what Max Planck has succeeded in doing, and it is the physical interpretation of this formula that is the starting point of the *quantum* theory.

In the isothermal enclosure thermal equilibrium exists between the matter of the walls and the empty space (or ether) in which there is black radiation. The temperature of the walls is due to the internal agitation of the molecules that constitute them; the temperature of the empty space is due to the stationary waves that exist between the walls. This thermal equilibrium between matter and radiation is realized by the mechanism previously described. The molecules of the walls of the enclosure in their kinetic agitation hit the free electrons existing in the walls. As a consequence of these shocks the electrons undergo accelerations that determine a radiation of all wave lengths, the electromagnetic energy of which is borrowed from the thermal energy of the molecular agitation: the temperature of the empty space is increased at the expense of that of the matter.

Reciprocally, and by a mechanism inversely symmetrical, the electrons, being subjected to the action of the radiations, vibrate in resonance with them. In consequence of the oscillations thus started they strike the neighboring molecules, the thermal agitation of which increases: the temperature of the matter is increased at the expense of that of the empty space. There is thermal equilibrium between the matter and the radiation, when there is equality between the exchanges of energy brought about by means of the electrons.

The aggregate of the material molecules, the electrons, and the various stationary waves that exist in the empty space, form a system in statistical equilibrium to which can be applied the laws of statistical mechanics, in particular the theorem, discovered by Maxwell, correctly demonstrated by Boltzmann, which sums up the properties of such systems. This is the law of the equipartition of energy. It states that in a thermal system the mean kinetic energy is divided equally among all the degrees of freedom of the system.

By *degrees of freedom* of a system are meant the different motions that it can assume consistently with the connections to which it is subject. For example, a material point can move along three axes, it has three degrees of freedom; a sphere can undergo a translation parallel to each of these three axes and a rotation about these axes, it has six degrees of freedom. The molecule of a monatomic gas, like argon, is supposed to have three degrees of freedom; a molecule of oxygen five; a triatomic molecule six, three degrees of translation and three degrees of rotation. If the law of equipartition is applied to a gas in thermal equilibrium, the *vis viva* which, on the average, each molecule will have, is proportional to the number of its degrees of freedom: if, in the statistical equilibrium, a molecule of argon possesses at a certain temperature the *vis viva* 3, a molecule of oxygen must possess the *vis viva* 5.

This law, which results from the Hamiltonian form of the equations of dynamics, must be applicable to the statistical system constituted by the material walls of the enclosure and the black radiation. It will be sufficient to count the number of the degrees of freedom of the system to predict, by this law, the most probable spectral composition of radiation in the state of thermal equilibrium, the realization of which is a necessary physical consequence of Carnot's principle. We have to consider on one hand the molecules of the material walls, on the other hand the various systems of stationary waves possible in the empty space. Let N be the number of the

material molecules contained in the walls of the
enclosure; the number of their degrees of freedom
being equal to 6, there will be, for the matter, as a
consequence of its discontinuity, only a finite number of degrees of freedom, namely $6N$. Let us consider on the other hand the empty space. It has an
infinite number of degrees of freedom, for there is an
infinite number of systems of possible stationary
waves, the wave-lengths of which lie between ∞ and
0. Hence, if the law of equipartition among all the
degrees of freedom is applied, the energy will be
found entirely in the empty space and none would
remain for the matter: equilibrium would cease to be
possible, or rather a single state of equilibrium only
would be possible, that in which the matter is at the
absolute zero. Moreover, the energy received by
the space must be equally apportioned among its
degrees of freedom, infinite in number. Whatever
might be the amount of the total energy, that assign-
ed to each degree of freedom would be zero, at least
if the quantity of energy at our disposal is not infinite, which has no physical meaning. This would
be no longer the case if it were assumed that the
length of the luminous oscillations can not descend
below a certain limit λ_0. This would bring us back
to assuming an ether and endowing it with structure.
In fact, what limits the periodic motions that can be
propagated in a given medium is the necessity that
their wave-lengths be appreciably larger than the
scale of the structure of the medium. The wave-
length of sounds given by a vibrating cord must be

greater than the mutual distance of the molecules
that constitute the cord or the sounds would not
exist. In the same way the seismic waves that
traverse a continent have reality only for an observer
whose horizon extends beyond the region of local
variations. Similarly, if a structure of the ether exists, we can no longer speak of infinitely small
wave-lengths in the mathematical sense of the word:
the smallest wave-lengths that it would be possible
to assume are those for which the magnitude λ is
near the number that measures the distance between
two ether molecules. The number of degrees of freedom of the ether would be limited in number, and
thermal equilibrium between matter and ether then
becomes possible.

This idea has the disadvantage of being based
on the hypothetical existence of the ether, and of an
ether endowed with discontinuous structure, which
is not readily conceived. But, moreover, it is
ineffective, for it leads to predictions at variance
with experiment. In fact, the number of systems
of stationary waves the wave-length of which lies
between the limits λ and $\lambda + d\lambda$ is greater the
smaller λ is. The result would be that the degrees
of freedom of the smallest wave-lengths would tend
to appropriate all the disposable energy, which
would be dissipated in extremely short radiations,
and this is contrary to experiment. Thus the
preceding hypothesis leads to a wrong law, formulated first by Lord Rayleigh and Jeans as a consequence of the equipartition of energy, that the

energy radiated for a given wave-length is proportional to the absolute temperature and varies inversely as the fourth power of the wave-length.

To escape from these difficulties, Planck has put forth a radical hypothesis, which has the merit of being free from the consideration of any hypothetical medium and considers nothing but the only positively accessible reality, energy. The formula that he has proposed for representing the distribution of energy in a black spectrum amounts to substituting discontinuous series of elements, the sum of which remains always finite, for an integral which occurs in the mathematical expression of Lord Rayleigh's law and which has the disadvantage of becoming infinite. He interprets this discontinuity, not by hypotheses on the structure of the medium in which the radiant energy moves, but by hypotheses on the absorption and emission of this energy. According to him the exchanges of energy between radiation and matter cannot take place in a continuous manner in any proportion whatever, for this introduces an infinite number of degrees of freedom, but they must take place in a discontinuous manner in definite proportions. The electric resonators (free electrons) through the agency of which these exchanges are realized can only absorb or emit radiant energy in a discontinuous manner, by sudden jumps, according to integral multiples of elementary quantities, indivisible energy atoms called *quanta*. These *quanta*, which fix the lower limit of intake and output of energy, are not the

same for all the resonators: they are inversely
as the wave-length (or the period of their oscillation)
and connected with the frequency ν according to
the relation

(21) $q = h\nu$

where h represents a universal constant. As a consequence of this hypothesis the resonators of short
period can absorb and emit energy only in large
mouthfuls, while the resonators of long period can
swallow and give it up in small morsels. It requires a large amount of disposable energy to rouse
a resonator of short period, so that the resonators
of this kind will have a chance to remain at rest,
especially if the temperature is low. By this means
the noxious role of the wave-lengths near zero, the
presence of which made the equilibrium impossible,
is eliminated. Thereby also the fact is explained
that there is relatively little light of short wave-
length in black radiation, which is in accord with
Planck's empirical formula. On the other hand,
the discontinuity in the intake and output of the
energy, which is progressively accentuated and becomes enormous for short wave-lengths, tends to
disappear in the region of large wave-lengths, where,
as the result, Lord Rayleigh's law is found again to
hold.

Let us consider our isothermal enclosure and see
what happens when the temperature is progressively raised, starting from the absolute zero. At
first the molecules are motionless and, as it were,

anchylosed by the frost, the thermal energy is zero, the resonators are mute. If the temperature is raised, the resonators whose *quantum* is the smallest will, because of the incipient molecular agitation, begin to vibrate, and the first radiations will appear in the infra-red which is far removed from the visible spectrum. Step by step the other resonators will begin to be agitated and the spectrum will be extended towards the luminous radiations, then towards the ultra-violet, in conformity with Planck's law and with what experiments reveal to us. The *quantum* increasing indefinitely as λ tends towards 0, the radiations corresponding to the limit will never appear.

Planck's theory seems thus to imply that the structure of energy is discontinuous. Every resonator can emit or absorb only a whole number of grains of energy. The value of this grain depends uniquely on the frequency v of this resonator and is proportional to it. There are atoms of energy, as there are atoms of electricity and atoms of matter formed of the first. However, while a hydrogen atom conserves its mass, whatever be the compound into which it enters, while the atom of electricity conserves its individuality through any vicissitude that it undergoes, it is not the same with the *quanta* of energy. If we have, for example, three *quanta* of energy on a resonator the wave-length of which is 3, and if this energy passes to a second resonator the wave-length of which is 5, it represents

no longer 3 but 5 *quanta* of energy. This is one
of the numerous difficulties of the theory.

One of the confirmations of the *quantum* theory
is drawn from the determination of Avogadro's
constant (the number of molecules contained in a
gram-molecule of a gas) which it provides. Two
constants figure in Planck's empirical formula: one
which expresses the discontinuity of the energy of
oscillation of resonators, called the *universal constant*
h, another which expresses the molecular discontinuity, called *Avogadro's constant N*. To determine
the numbers *h* and *N* it suffices to use two good
measurements of the emissive power (the quantity
of energy that leaves the opening of the enclosure
per second) for different values of the wave length
λ and the temperature T. The value of h is then
found to be

$$h = 6.2 \times 10^{-27}$$

and this leads to the value of N

$$N = 64 \times 10^{22}$$

This number agrees quite closely with the mean of
the numbers obtained by the methods considered
most reliable, and this coincidence is the more surprising as the number of molecules is deduced from
measurements made on radiant heat.

A second confirmation, due to Einstein and to
Nernst, is drawn from the study of the specific heats
of solids at low temperature. It is known that
the specific heats of solids decrease rapidly when
the temperature is lowered. Thus, for diamond

at the temperature of liquid hydrogen, the specific
heat is reduced to about one-seventh of what it is
at ordinary temperatures. Everything takes place
as if the molecules lost degrees of freedom in cooling, as if their joints became anchylosed from the
effect of the frost. This is contrary to the theorem of the equipartition of energy, which provides
a method of calculating the atomic heats of solids
and of deducing Dulong and Petit's law, according to which the atomic heats are equal for all bodies
and independent of the temperature.

Planck defined the specific heat of a single
resonator as the increase, for one degree centigrade,
of the mean energy, reduced to calories, that a
resonator of given frequency must have at a definite
temperature to be in equilibrium with black radiation. Einstein introduces the following simpli-
fying hypotheses. He considers the solid bodies as
possessing only a single kind of resonators and only
one resonator per molecule; the product of Avogadro's constant by the specific heat of a single
resonator then gives him the specific heat, referred
to a gram-molecule, of the solid considered. According to this formula the specific heat of a solid
hardly varies with the change of temperature
at high temperatures; but at low temperatures it
decreases rapidly and tends towards zero when the
absolute zero is approached. Nernst complicates
Einstein's hypothesis a little by adding to the latter's unique system of resonators other resonators
tuned to the octave and so succeeds in obtaining

a surprising coincidence with the numbers obtained experimentally by himself and his pupils, for a very large number of bodies at temperatures that extend from the ordinary temperatures down to the lowest temperatures that Kammerlingh-Onnes has been able to obtain in his cryogenic laboratory at Leyden. This theory amounts to the following: if a solid is regarded as an aggregate of atoms or of molecules oscillating about an equilibrium position, the energy of each oscillator thus realized must be, as in the case of Planck's electric oscillators, an integral multiple of $h\nu$. The diminution of the specific heat of the solid at low temperatures is then readily explained. When the temperature decreases, the supply of disposable energy offered to each of the material oscillators falls below the *quantum* of a large number of them; instead of vibrating a little they cease to vibrate at all, so that the total energy diminishes more rapidly than in the old theories. On the contrary, at high temperatures, the *quantum* $h\nu$ becomes so small that we again get Dulong and Petit's law, obtained by starting from the law of the equipartition of energy.

19. THE STRUCTURE OF RADIATION.

The study of black radiation and of specific heats at low temperatures has led Planck, and after him, Einstein to the idea that the exchanges between radiation and matter do not take place in a continuous manner, but by discrete elements, by *quanta* of energy. What becomes of these elements when once set free by matter? For the old theory of free radiation considered as formed by indefinitely divisible and expansible spherical waves propagated by a continuous hypothetical medium, there is now substituted the idea of a radiation projected in space void of matter in the form of distinct elementary units, which implies a discontinuous distribution of energy in the front of the luminous transverse waves. This is an unexpected return to the emission theory, rendering the consideration of an ether serving as the "I" to the verb "undulate" superfluous.

Among the phenomena, the interpretation of which suggests this point of view, we have Hertz's phenomenon and the production of rays by shocks of secondary cathode rays and by X-rays.

Hertz's phenomenon, or the photo-electric effect, denotes the property that bodies have, in particular metals, of emitting cathode rays under the action of light. The electric forces present in the absorbed luminous waves set the electrons stored in the metal in motion, and some of them are, because of the accelerations that they undergo, projected out. The photo-electric current, which results therefrom, depends on two factors, the number of the electrons emitted per unit time and the velocity with which these electrons are emitted. If the intensity of the light is increased without a change in its wave-length, the number of the electrons emitted increases proportionally, but the velocity of emission remains the same. If, on the other hand, at constant luminous intensity a change is made in the frequency ν of the exciting light, the velocity of emission seems to increase proportionally to the frequency, and consequently inversely as the wave length. The velocity with which the electrons are emitted certainly did not belong to them before they had undergone the action of the light, as experiment shows that this velocity is very much higher than that of the thermal agitation. If the velocity of the electrons is borrowed from the light, how are we then to understand that it is the same for a very intense light and a very feeble light? How, above all, explain that it varies inversely as the wave length?

Stark[1] has remarked that these difficulties disappear if the *quantum* hypothesis is assumed. The light that meets the metal is not of homogeneous structure, it consists of isolated elements each possessing the energy $h\nu$. When one of these elements strikes the metal the energy that it contains can pass to one of the electrons present in the metal. On account of this fact the electron acquires a velocity that can much surpass the velocity of thermal agitation and leads to its projection from the metal. As the elements of energy are greater in proportion as the frequency ν is higher, ultra-violet light will be much more effective than visible light. Thus not only is light absorbed and emitted solely in the form of *quanta*, but, even when it is freely pro-

[1] Stark, Phys. Zeitschr., 1909; Principien der Atomdynamik. Leipzig, 1910.

pagated in empty space, it seems to consist of the projection of discrete units.

J. J. Thomson[1] has arrived at similar conclusions by studying the photo-electric effect on gaseous bodies. Ultra-violet light of very short wavelength ionizes gases, that is, it breaks up the gaseous molecules by setting electrons free. The work necessary for this rupture can be determined. It is thus found that if, in conformity with Maxwell's ideas, ultra-violet light is thought of as formed by a train of homogeneous waves propagated without inequalities of structure, the energy carried through a unit section is not sufficient to account for the work of ionization produced by this pencil. To do this a discontinuous front must be assumed, so that there are spots of light and darkness. At places where the energy is accumulated its density is sufficient to ionize the molecules. The wave front possesses a structure, and the wave train, taken as a whole, is not formed of a regular and uninterrupted flow of energy, but of discrete elements of energy separated by considerable gaps. This is the old emission theory reappearing in a very modified form.

Einstein has supplemented Stark's ideas in a way that has made a new experimental verification of the *quantum* theory possible. He assumes that an electron set free by a luminous ray of frequency ν can bring into action only an amount of energy equal to Planck's energy element. Consequently

[1] J. J. Thomson, Proceed. Camb. Phil. Soc., 1908.

we have, according to the principle of the conservation of energy, the equality

$$\tfrac{1}{2}mv^2 = h\nu - \Psi$$

where Ψ denotes the work necessary to separate the electron from the atom from which it comes. This formula is capable of direct verification, if we take into account the fact that the velocity v of the photo-electric electrons can be determined experimentally by the measurement of the condenser potential, $V + V_0$, capable of reducing the current to zero, by reducing the velocity v to zero. We have

$$e(V + V_0) = \tfrac{1}{2}mv^2$$

where V denotes the potential difference applied, and V_o the contact potential difference that always exists between two different metals, and these two differences have to be determined separately. Among the most satisfactory verifications of this formula, the recent experiments of Millikan[1] must be quoted. Working in a vacuum on freshly prepared surfaces of sodium, potassium, and lithium, this physicist has employed the entire range of wave-lengths from the middle of the visible spectrum to the ultra-violet of the mercury arc. His results confirm Einstein's theory. In particular, the linear relations that he obtains between V and ν make possible the determination of a constant angular coefficient practically equal to Planck's constant.

[1] R. A. Millikan, A Direct Photoelectric Determination of Phanck's "h." Phys. Rev. Ser. 2, vol. 7, p. 355.

Lorentz[1] has shown, not only that the corpuscular idea of light results from Hertz's phenomenon, but that it alone permits the explanation of the absorption of radiation by fixed *quanta*, multiples of $h\nu$, in conformity with the views of Planck. When an incident wave, encountering the molecules of an isotropic body, sets the electrons in it into vibration, the absorption of light is the more intense the nearer its frequency is to the natural frequency of these electrons. In the case of perfect resonance the electron borrows from the luminous wave the maximum amount of energy, and, by taking the case of the solar light and allowing for the forces of damping and friction to which the electron is subject, this maximum energy is found by calculation to be a little more than twice the atom of energy required by Planck. A similar amount of light, therefore, can be absorbed by the electrons. But the ordinary sources have intensities a million times weaker than that of the Sun. If the energy that they emit were uniformly distributed over the surface of a sphere having the source as center, it could communicate to the electron only a very small fraction of its absorbable quantum. This energy, therefore, must be unequally concentrated in different directions, and the sphere must be traversed by the flow of energy in a certain number of discontinuous spots, very widely spaced, to provide condensations of energy compatible with the mechanism of absorption conceived by Planck.

[1] H. A. Lorentz, Phys. Zeitschr., 1910.

The *quantum* theory, it is hardly necessary to say, encounters otherwise many difficulties. One is the existence of interferences at large differences of path, which seems to imply that the luminous *quanta* extend over lengths equal to several thousand times their wave-length. There are others which, however, yield to a deeper analysis. The idea that a point source does not radiate symmetrically in all directions may seem incompatible with experiment. But the physical point sources contain a number of oscillators sufficiently great to wipe out any trace of individual discontinuity. The physical symmetry, which experiment leads us to attribute to luminous waves, results from a mean effect and is not in contradiction to the possibility of an elementary radiation, endowed with structure. This has been shown by J. J. Thomson[1] by supposing that the electrified particles producing the light do not emit lines of force uniformly in all directions, but only, for example, in two solid angles opposed at the vertex and of relatively restricted aperture. Starting from this hypothesis, he has demonstrated that the principal properties of the electron would still belong to those particles, the electromagnetic field of which is concentrated in privileged directions. They would have inertia of electromagnetic origin and would emit radiation on every modification of their state of motion; but the radiated waves of acceleration, light, if we are concerned with periodic waves,

[1] J. J. Thomson, Phil. Mag., 1910, p. 301.

would be concentrated in certain directions and extraordinarily rarified or absent in others. The radiation emitted by an oscillator would possess a particular structure, which would reappear in part in the emission of real sources and produce the irregularities of the wave front suggested by Hertz's phenomenon. Nothing essential would be changed in the usual laws of electromagnetism, and they would continue applicable to material complexes enclosing a large number of particles distributed according to the law of chance.

The *quantum* theory is far from having the dogmatic certainty of the atomic theory or the electronic theory. The atoms, for example, have become a physical reality: they are counted, they are weighed, their radius of action and their mean velocity are determined, they are seen in the form of scintillations with the aid of the spinthariscope of Crookes, and C. T. R. Wilson, by using their property of becoming the centers of condensation in supersaturated water vapor, has succeeded in photographing the atomic trajectory of the α particles, which are ions of helium, and even the corpuscular trajectory of the β rays, which are electrons. No device, up to the present, has enabled us to subject the *quanta* of radiant energy to a control capable of transforming them from theoretical concepts to experimental concepts. One, however, can be suggested, drawn from the marvelous properties of selenium.[1]

[1] Cf. Fournier d'Albe, The Future of Selenium (Scientia, 1917, p. 165–191).

When one of the conductors in an electric circuit consists of selenium, a current is produced in this circuit, as soon as it is subjected to a luminous flash. The conductivity thus produced from an instantaneous exposure of the selenium to light is proportional to the incident energy. This property makes it possible to detect, electrically, flashes that are invisible to the naked eye. By applying, for example, an electromotive force of 1 volt, an easily perceptible current of 10^{-12} ampere could be obtained with a cell of selenium of 100 sq. cm., with a short flash of 10^{-9} lux, corresponding to a star of 8th or 9th magnitude, which is quite invisible to the naked eye. With a selenium cell of the same surface as the pupil of the eye, a star of the 6th order, which is the extreme limit of our vision, could be detected electrically under the same conditions. But, since we are able to measure currents of 10^{-15} ampere, the sensitiveness of selenium obviously surpasses that of the eye. Now the *quantum* theory permits the calculation of the size and number of the grains of energy absorbed by a given surface, subject to a radiation of definite wave-length and intensity. It is found that the human eye absorbs 360 *quanta* per second, when it receives light coming from the visible star of least luminosity. These *quanta* are, in the most favorable case, 20 times too numerous to be perceived as distinct flashes. We can then hope to count them electrically, aided by the selenium, if we can succeed in detecting sufficiently feeble currents.   The selenium cell would thus render to us the same service as the zinc sulfide screen of Crookes' spinthariscope; it would enable us to place the existence of the *quanta* on an experimental basis.

20. THE PHYSICS OF THE DISCONTINUOUS.

Modern discoveries lead to the assumption of the discontinuity of matter, of electricity, of radiant energy, and of exchanges of energy.   Henri Poincaré has shown the consequences that result therefrom for mathematical physics and natural philosophy.   We must give up expressing the laws of the phenomena in the form of differential equations in all cases in which the large number of elements which come into play does not suffice to wipe out entirely the influence of the individual discontinuities.   Contrary to the ancient adage *natura non facit saltus*, it becomes apparent that the universe varies by sudden jumps and not be imperceptible degrees.   A physical system is capable of only a finite number of distinct states, and this introduces discontinuity into the distribution even of probabilities.   Since between two different and immediately consecutive states the world remains motionless, time is suspended so that time itself is discontinuous: there is an atom of time.

The controversy between infinitists and finitists, idealists and empiricists, cantorians and pragmatists, to follow current terminology, is, on this basis, settled in favor of the latter.   The world does not glide smoothly down a gentle slope, imperceptibly inclined, of the course of events according to Leibniz, but it descends by steps of events as conceived by Evellin.[1] In the external reality there are no aggregates of elements having the attributes of a continuum ,but only aggregates that can be counted, the attributes of which are only intelligible for the mind, because it can define all their elements. *Nature following mathematics is reduced to arithmetic.* It becomes comprehensible, for if our imagination does not show us anything except in the form of spatial intuition, which is that of a physical continuum, our mind, by virtue of the minimum perceptible increment of excitation, can comprehend only the countable and the discrete. Just as continuity in mathematics appears more and more as a transitory tool, the utility of which at present is not negligible, but which must be regarded as a means of studying the countable aggregates that constitute the only analytical reality that is accessible, so too physical continuity, which lends itself well to applications of the calculus of partial derivatives, will always appear legitimate, as a first approximation, for the order of magnitude of ensemble effects of systems the elements of which are finite in number but sufficiently numerous to wipe out the individual discontinuities by virtue of perfect mixing and the laws of chance. Nevertheless it should not make us forget that the simple elements of things manifest an essential discontinuity, which must reappear in the equations that translate their individual behavior

[1] A contemporary French philosopher. (Tr.)

and which alone can account, in the hypothesis of the absolute determinism, for the fluctuations of the ensemble phenomena to which they give rise.

On mathematical physics there is hereafter imposed a new task, namely, to establish a bond:—

Between the ensemble phenomena which our observations reveal, and the elementary phenomena of which they are a statistical resultant;

Between the physical quantities directly accessible to our measuring instruments, which take in at once, as the sum or average of the individual quantities, so many elements that they can be practically treated as continuous; these quantities being in themselves, or individually, essentially discontinuous;

Between the properties of the discrete grains (molecules, atoms and electrons) and the structures they form by aggregation.

The differential and integral calculus, used to translate analytically the idea of continuity, is appropriate for the study of the systems which alone are directly perceptible, and which are composed of a very great number of elements. The calculus of probabilities is appropriate for the study of the relations between the real world of elements of discontinuous structure and the apparent world of continuous phenomena; between the individual laws, which govern these elements taken separately, and the laws of large numbers, which govern the mixed appearances to which their incessant mixing gives rise.

10

By correlation, the concept of explanation becomes transformed in natural philosophy. In former days a physical phenomenon was explained by reducing it to the principles of classical mechanics, by giving to its laws the form impressed by Lagrange on the equations of dynamics. To explain a phenomenon today is to give a statistical explanation, by regarding it as the resultant of a very large number of underlying phenomena governed by the laws of chance. Maxwell's attempt in his *Treatise on Electricity and Magnetism* is an example of an explanation of the first type. The statistical interpretation of Carnot's principle, of the exponential law of the spontaneous destruction of radioactive substances, and of the law of mass action are examples of the second type.

Thus not only do the most fundamental categories of our mind, those of space, of time, and of causality, pass through an evolution with the progress of science, but the same holds even for the concept of intelligibility. To explain a phenomenon is, for primitive man, to interpret it anthropomorphically by a supernatural agent endowed with psychological life in his own image; for a scholastic it is to explain it by ultimate causes; for Bacon to explain it by efficient causes; for Maxwell it is to deduce it from the principles of mechanics; for Gibbs and Boltzmann it is to account for it by the calculus of probabilities, by starting from a system of elements subject to given conditions. Human reason is not "une et entière en un chacun" as

Descartes taught. It varies with the abstract or concrete nature of our thought, and in proportion as, on contact with experimental facts, the adaptation of our mind to nature becomes progressively realized.

CHAPTER VIII

CONCLUSION

21. CONCLUSION.

The discoveries of modern physics have led physicists to two quite distinct conceptions of the universe.

The first can be illustrated by the suggestive name of the *dematerialization of matter*. It consists in reducing matter to being only the locus of singular points of torsion, of condensation, or even of destruction of a medium endowed with inertia and mechanical properties, the dielectric ether of Faraday and Maxwell. From this point of view an electron is a simple cell in the ether, behaving like a projectile moving forward in a perfect fluid having no viscosity. On the sides of the projectile a cushion of fluid is formed and behind it a zone of deep eddies: its own inertia is increased by all the inertia of the wave system thus created which follows it. At start a part of the muzzle energy is expended in overcoming the inertia of the fluid displaced at the same time as the inertia of the moving body itself; but, when the motion has been once acquired, it perpetuates itself without resistance, since the body carries its wave system with it. The electron has no inertia of its own, but it can not be displaced without entraining

148

the surrounding ether coupled to its lines of force, and its inertia results from that of the ether thus disturbed which forms its electromagnetic wave system. Matter resolves itself into cells in the ether and the ether gains in substantiality and in reality all that matter loses.

This idea results from Faraday's work, which has brought to light the importance of the medium that surrounds conductors and magnets in electric and magnetic phenomena. The attention of physicists has been directed to the study of electric and magnetic fields in which energy is concentrated, matter serving only as a support of these fields. But instead of conceiving them as substantial realities existing in an independent manner, it is claimed that they are explained by the mechanical states of this hypothetical medium, the ether, electric energy being merely the potential energy of its deformations and magnetic energy merely the kinetic energy of its displacements. The ether becomes the active medium in which the transfers and the transformations of energy are governed by the equations of the electromagnetic field of Maxwell and Hertz, and thus matter is gradually stripped of all its physical contents and they are referred to the ether, which becomes the only reality that continues to exist.

Going still further Dr. Gustave Le Bon[1] has

[1] Dr. G. Le Bon, L'évolution de la matière. On the relations and the differences between this theory and that of Einstein see L. Rougier, L'inertie de l'énergie (Revue scientifique, October 13–20, 1917).

developed the idea of the evolution of matter. Let us imagine that the electrons, into which in the end the molecular structures that constitute bodies are resolved, are due to vortices in a universal fluid analogous to Lord Kelvin's gyrostatic ether. Instead of regarding these vortices as indestructible in conformity with the hydrodynamic equations of Cauchy and Helmholtz, let us imagine, quite gratuitously, that they vanish at length in the original fluid, like a waterspout in the ocean, in the form of liquid waves, owing to the gradual retardation of their velocity. The rotatory energy of the electron will be transformed into radiant energy, which, sweeping through all space, will be lost at infinity. Thus matter is resolved into electrons, which themselves vanish in etherized undulations, so that there is a final loss of matter and an uncompensated dissipation of energy. For the universal principle of invariance which the Ionic natural philosophers placed at the basis of natural philosophy and which assured its intelligibility, namely, "Nothing is created, nothing is lost" one must now substitute the contrary principle: "Nothing is created, everything is lost." The world marches toward a final bankruptcy, and the ether, of which it has been asserted in vain that it is the matrix of the worlds, is revealed as being their final tomb. Thus Dr. Gustave Le Bon has been, in sumptuous prose, the Zarathustra announcing, after the death of the gods, the twilight of their creation.

These ideas encounter insurmountable difficulties.

The ether is revealed as endowed with contradictory mechanical properties, and the attempts at an explanation of electromagnetic phenomena starting from it have all suffered shipwreck. If it exists it can not be completely entrained by matter, as Fizeau's experiment proves; it can not be partially entrained by matter, as the principle of action and reaction demonstrates; that it can not be motionless, is implied by the relativity principle. Less tangible than Proteus, it remains only to declare it defunct without estate.

To these antinomies the theory of Dr. Gustave Le Bon adds new difficulties. The hydrodyn mic equations of Cauchy and Helmholtz show that vortices, once started in a perfect, homogeneous, and incompressible fluid, are eternal. The electronic theory of radiation connects its appearance with the presence of electrons, which play the part of agents present in the transformation of different forms of energy into radiation and permit these transformations, while themselves remaining, like catalizing agents, unaltered. Their disappearance would entail therefore that of radiation itself, in which they are supposed to vanish. Lastly, if the principle of invariance of the Ionic natural philosophers is abandoned, the very possibility of science is put at stake. Science, being the search for the laws of nature, that is to say, the invariants of all events in the universe, would remain valid, to a first approximation, only for systems in which the disintegration of matter and the radiation of energy to infinity are practically insignificant; and the metaphysical problem would arise of understanding how the universe, if it has not had a beginning, has not yet finally vanished, milleniums ago, in the "motionless and sleeping" ether.

Abandoning the ether, we are led to an entirely different theory: that of the *materialization of energy*. Energy emerges from the phantom realm of imponderables to take substance, like the shades of the Elysian fields evoked by Ulysses on the Cimmerian river. It appears as endowed with inertia, with weight and structure and manifests itself in two forms: one is called, by virtue of long prescription, *matter;* the other, *radiation*.

Matter is characterized by its structure, that is, by the number and nature of the electrons and perhaps the positive remainders that constitute it, and also by its property of moving with velocities, relatively to a reference system, ranging from 0 to V. We know nothing of it, as stated by Ostwald, except its energetic effects: the electric field of the electron at rest, the magnetic field of the electron in motion, the gravitational field of the molecular structures formed by electronic architecture, the kinetic effects produced by their *vis viva*. The electron is revealed as a grain of resinous (negative) electricity, so that matter is only a form of energy, enormously accumulated in a narrowly circumscribed region of space. It does not, on that account, as in the preceding theory, lose the reality and the substantial characteristics which external perception and common sense have agreed until now to attribute to it, since energy, which is its essence, is endowed with mass, weight, and structure.

Radiation is a form of energy which no longer appears as propagated in the shape of continuous waves by a hypothetical medium, but as expelled in the form of discrete units in space free of matter with the uniform velocity of light. It also is endowed with inertia, weight, and structure. Its possession of fundamental properties in common with matter permits the explanation of its action on the latter. Luminous radiation, representing a certain momentum, can be strictly assimilated to a material projectile. It is because of this that, by virtue of the principle of action and reaction, it exerts a repulsion on the material source that emits it and a propulsion on a material obstacle that absorbs it. The ancient metaphysical problem of the action of the imponderable on the ponderable, of force on matter, which arose in its most modern and most urgent form in connection with the pressure of radiation, disappears henceforth as a pseudo-problem.

BIBLIOGRAPHY

1. The Relativity Principle

H. A. LORENTZ, A. EINSTEIN, H. MINKOWSKI, Eine Sammlung Abhandlungen, Teubner, 1913.

LAUE, Das Relativitätsprinzip, Teubner, 1911.

P. LANGEVIN, Le temps, l'espace et la causalité dans la Physique moderne (Bulletin de la Société française de Philosophie, t. XII, 1912).

2. Electromagnetic Dynamics

H. POINCARÉ, La théorie de Lorentz et le principe de réaction (Archives néerlandaises des Sciences exactes et naturelles, 2e série, t. V, 1900).

H. POINCARÉ, Sur la dynamique de l'électron (Rendiconti del Circolo matematico di Palermo, t. XXI, 1905).

P. LANGEVIN, Les grains d'électricité et la dynamique électromagnétique (Les idées modernes sur la constitution de la matière, Paris, 1913).

3. The Electronic Theory of Matter

P. LANGEVIN ET M. DE BROGLIE, Les quantités élémentaires d'électricité, ions, électrons, corpuscules, 2 vol., Paris, 1905.

Les idées modernes sur la constitution de la matière (Mémoires de la Société française de Physique, Paris, 1913).

N. R. CAMPBELL, Modern Electrical Theory, Cambridge, 1913.

4. The Inertia of Energy

A. EINSTEIN, Annalen der Physik, vol. XVIII, 1905.

P. LANGEVIN, L'inertie de l'énergie et ses conséquences (Journal de Physique, juillet 1913).

5. The Weight of Energy and the Theory of Gravitation

A. EINSTEIN, Annalen der Physik, vol. XXXV, 1911; Ibid., vol. XXXVII, 1912; Physikalische Zeitschrift, vol. XIV, 1913.

A. EINSTEIN ET GROSSMANN, Bases physiques d'une théorie de la gravitation (Archives des Sciences physiques et naturelles, t. XXXVII, 1914).

A. EINSTEIN, Die formalen Grundlagen der allgemeinen Relativitätstheorie (Sitzungsberichte der königlich preussischen Akademie der Wissenschaften, vol. XLI, 1914).

A. EINSTEIN, Erklärung der Perihelbewegung des Merkur aus der allgemeinen Relativitätstheorie (Sitzungsberichte der königlich preussischen Akademie der Wissenschaften, vol. XLII, 1915).

H. A. LORENTZ, La gravitation (Scientia, 1914).

ED. GUILLAUME, Les bases de la Physique moderne (Archives des Sciences physiques et naturelles, 1916).

L. BLOCH, Relativité et gravitation, d'après lés théories récentes (Revue générale des Sciences pures et appliquées, 15 décembre 1917).

L. Bloch, Sur les théories de la gravitation (Annales de Physique, janvier-février 1918).

6. The Structure of Energy

P. Langevin et M. de Broglie, La théorie du rayonnement et les quanta, Paris, 1912.

H. Poincaré, Sur la théorie des quanta (Journal de Physique, janvier 1912).

H. Poincaré, L'hypothèse des quanta (Revue rose, 24 février 1912).

P. Langevin, La physique du discontinu (Les progrès de la physique moléculaire, Paris, 1914).

M. Planck, Die physikalische Struktur des Phasenraumes (Annalen der Physik, vol. L, 1916).

P. S. Epstein, Zur Quantentheorie (Annalen der Physik, vol. LI, 1916).

A. Einstein, Zur Quantentheorie der Strahlung (Physikalische Zeitschrift, vol. XVIII, 1917).

Other Works by L. Rougier:

Les Paralogismes de Rationalisme, Paris, 1920. La philosophie géometrique de Henri Poincaré, Paris, 1920. En marge de Curie et d'Einstein, Paris, 1920. Théories formelles et logique déductive, Paris, 1921.

INDEX OF NAMES

157

www.ingramcontent.com/pod-product-compliance
Lightning Source LLC
Chambersburg PA
CBHW071716170526
45165CB00005B/2035